Forensic Digital Image Processing

Optimization of Impression Evidence

Forensic Digital Image Processing

Optimization of Impression Evidence

Brian E. Dalrymple
E. Jill Smith

CRC Press is an imprint of the
Taylor & Francis Group, an **informa** business

CRC Press
Taylor & Francis Group
6000 Broken Sound Parkway NW, Suite 300
Boca Raton, FL 33487-2742

Printed on acid-free paper

International Standard Book Number-13: 978-1-4987-4343-3 (Hardback)

Library of Congress Cataloging-in-Publication Data

Names: Dalrymple, Brian, author. | Smith, Jill, 1968- author
Title: Forensic digital image processing : optimization of impression evidence / Brian Dalrymple and Jill Smith.
Description: Boca Raton, FL : CRC Press, [2018] | Includes bibliographical references and index.
Identifiers: LCCN 2017054403| ISBN 9781498743433 (hardback : alk. paper) | ISBN 9781351112239 (ebook)
Subjects: LCSH: Legal photography. | Image processing--Digital techniques. | Forensic sciences.
Classification: LCC TR822 .D35 2018 | DDC 770.2/436325--dc23
LC record available at https://lccn.loc.gov/2017054403

Visit the Taylor & Francis Web site at
http://www.taylorandfrancis.com

and the CRC Press Web site at
http://www.crcpress.com

Please visit the eResources at: www.crcpress.com/9781498743433

Printed and bound in the United States of America by Sheridan

"Begin at the beginning," the King said, very gravely,
"and go on till you come to the end: then stop."

Lewis Carroll
Alice in Wonderland

Contents

Foreword

Henri Cartier-Bresson, a French-born, humanist photographer, once said, "Your first 10,000 photographs are your worst." For a time, I thought he was talking about me when I first started photographing evidence, but I'll have you know, those 10,000 photographs helped keep Polaroid in business!

As I look back at the evolution of forensic digital imaging over the past two-plus decades, I am amazed not just by the advancement of digital technologies, but by the speed at which it has evolved! I remember when crime scene photographs could be shown in court only in black and white, for fear of influencing the jury with inflammatory photographs of a bloody crime scene in full color.

But even after color photography was accepted in the courts, many law enforcement agencies continued to photograph evidence, especially latent prints, using traditional black and white photography until the late 1990s. I truly believe that most if not all Automated Fingerprint Identification Systems (AFIS) in the world still have unsolved latent prints in their databases that were developed with ninhydrin and photographed using a green filter and black and white photography, and the unremoved background prevents these unsolved latent prints from being identified.

Forensic digital imaging experienced a few challenges and setbacks in the early days. For example, Wisconsin was one of the last states to allow the use of digital imaging in the criminal justice system in the United States. On October 13, 2003, the Wisconsin State Legislature enacted Assembly Bill 584, which read in part: "[t]his bill prohibits the introduction of a photograph [in court] ... if that photograph ... is created or stored by data in the form of numerical digits." Section 3. 910.01 (2) of the statute was also amended to read: "[p]hotographs include still photographs, X-ray films, motion pictures, and any digital representation." This law was not rescinded until 2007. In the meantime, it probably was a good thing that the Wisconsin legislature did not have a thorough understanding of "digital representations" or perhaps someone forgot to tell them about a dirty little secret: every fingerprint on every 10-print card as well as every latent print entered into the Wisconsin AFIS—the central repository for fingerprint records in the state since AFIS was installed in 1993—was "stored by data in the form of numerical digits." (Now isn't that a kick in the pants!)

But, thanks to the untiring and enduring efforts of Brian Dalrymple and Jill Smith, the advancement and awareness of the development, detection, digital capture, and digital processing continues to progress at an even faster pace. Through the sharing of their knowledge, experience, and expertise in this book, they continue to enhance the tools and the techniques that are growing ever more crucial to identifying criminals and documenting crime scenes as digital imaging technologies continue to progress.

Anyone who is interested in, involved in, or even imagining they want to be interested in or involved in forensic digital imaging must have this book! As the old cliché goes, "there are all kinds of books on bookshelves in Hollywood because the scripts didn't capture the characters," but this book needs to be on every bookshelf in every law enforcement agency because it definitely will help capture the criminals!

David "Ski" Witzke
Vice President
Program Management
Foray Technologies

Acknowledgments

- Joseph Almog, for his constant generosity and consultation support.
- Erik Berg, for his extensive knowledge, his contributions to the digital discipline, and his sharing of crucial information related to the landmark investigations in which digital imaging has played such a pivotal role.
- Jonathan Cipryk, for sharing his cutting-edge knowledge of digital technology.
- Ed German, for his introduction to digital image processing and his memories.
- Darryl Hawke, for his expert co-teaching and IT support.
- Brad Joice, for progressively moving us forward and providing further training in digital imaging, as well as his part in moving the whole province forward with his involvement in the Ontario Forensic Investigators Association (OFIA), www.ofia.ca.
- John Jones from the Organization of Scientific Area Committees (OSAC), for providing us with a high-resolution image of their organization chart and permission to use it in Chapter 2.
- Dave Juck, for providing the training and tools required to enable our forensic imaging unit to grow and for providing the support and encouragement for *me* to grow (Jill).
- Carl Kriigel, for reviewing Scientific Working Group on Imaging Technology (SWGIT) and OSAC information and providing input and insight into Chapter 2.
- Nancy Merriman, Manager, Communications Support Unit, Ontario Provincial Police, for her gracious assistance in the permissions process.
- Myriam Nafte, for her generous support and valuable consultation.
- John Norman, OPP Forensic Identification Services (ret.), for his knowledge and assistance with historic case information.
- Pam Ringer, formerly of Hunter Graphics, for her warm friendship and generous assistance in the compilation of data from cases of past decades.
- Dana Rosenthal, for her artistic passion and kindness in the use of images.

- Bill Watling, for his friendship, enthusiastic training, his knowledge, and support.
- Dave Witzke, for his inspiration as a true pioneer in digital imaging, for providing the Foreword to this book, and for always being an invaluable resource of information and a refreshing source of positive energy.
- York Regional Police, for ultimately investing in the training and education, hardware, and software to make us competitive in this field and for supporting me (Jill) in the contributions of this book, as some images were provided from the casework of our police service.

Authors

Brian E. Dalrymple, CLPE, began a career in identification in 1971 with the Ontario Provincial Police, Forensic Identification Services, Canada. In 1977, he co-developed the technique of evidence detection with argon lasers through inherent fluorescence and became the first in the world to operate a laser for evidence detection in criminal cases. The extensions of this technology are now in global use and have provided pivotal evidence in hundreds of major investigations. In 1991, as associate section head, he introduced the first computer evidence system to Canada and later became the first Canadian to tender expert evidence in this emerging technology. In 1992, he was promoted to manager, Forensic Identification Services, a position he held until his retirement in 1999. He initiated and co-wrote the body examination protocol for the Province of Ontario regarding the examination of murder victims for fingerprints on skin. Dalrymple has taught extensively in North America, China, the Middle East, and Australia. He is the recipient of the John Dondero Award (1980, IAI), the Award of Merit (1980, Institute of Applied Science), the Foster Award (1982, the Canadian Identification Society), and the Lewis Minshall Award (1984, the Fingerprint Society, UK). Dalrymple has been a contract instructor for the Ontario Police College for more than a decade and provides forensic consulting for police agencies, attorneys, and the corporate sector. With Ron Smith & Associates, he is a consultant and staff instructor working as part of a team of forensic experts. He is an adjunct professor at Laurentian University in the Forensic Science Department.

E. Jill Smith is a forensic imaging specialist with the York Regional Police in Ontario, Canada, working in forensic digital image enhancement of impression evidence intended for analysis since 1999. She has been published several times in the *Journal of Forensic Identification*, published by the International Association for Identification (IAI), and served on the IAI Forensic Photography & Imaging Certification Board from 2013 to 2017. Smith has lectured and instructed at the Ontario Police College, as well as with Brian Dalrymple across the United States, for over 10 years.

Introduction

Chapter 1 explores the emergence of forensic digital image processing from the viewpoints of the authors, the contributions of pioneers in this discipline, again from the personal experience of the authors, and the gradual improvement and acceptance of the science over the past four decades. Image processing of evidence began in the 1970s with digital enhancement of analog images, long before digital technology displaced film. Finally, we review the radical changes in the job description of forensic photographers.

Chapter 2 delves into the issues of image integrity and authentication dictated by the rapid transition from film negatives to digital photography, examining the professional entities that have guided and enabled the transition process, creating solid protocols for the secure and trustworthy application of these procedures. The differences and the safeguards between professional forensic image optimization and the manipulation of images in movies, television, and advertising are compared and contrasted.

In Chapter 3, the different strategies between analog (film) and digital enhancement of images, both pre-capture and post-capture, are discussed. A progression of techniques exploiting color theory, modes, and channels may be used to optimize signal-to-noise ratio in images.

Chapter 4 features one of the greatest assets of digital image technology—the ability to combine multiple images of the same subject to create a final blended image that displays the desired evidence, be it a fingerprint or footwear impression, in optimum focus and with substrate interference diminished or removed entirely. Image subtraction, focus stacking, and high dynamic range are presented and demonstrated.

Chapter 5 presents and explores fast Fourier transform, one of the most powerful and underutilized strategies for noise removal in images, converting an image from the spatial to the periodic domain, where editing of pattern interference can be easily completed. Basic theory and diagnosis of the noise signatures revealed in the transform are examined.

"Remove the noise to the degree possible before adjusting contrast" has become something of a mantra in the world of digital image processing. In Chapter 6, the commonly used adjustment tools for optimizing contrast are illustrated and discussed.

Chapter 7 summarizes the four general enhancement steps necessary for image enhancement, followed by a series of practical exercises that review the strategies covered in previous chapters.

Chapter 8 explores the history of digital imaging and techniques in court, starting with the first unsuccessful attempt to introduce it in 1972, and its emergence as a trusted and accepted science across North America. Selected pivotal cases, challenges, and successes in courts of all levels are outlined.

Note that images used for many of the exercises in this book are available for download, so you can work along. Please visit www.crcpress.com/9781498743433.

History of Forensic Digital Enhancement

1

The secret of getting ahead is getting started.

Mark Twain

The goal of this chapter is to give context to the emergence of digital image processing in the forensic science domain and to explore the reasons why film is no longer the default recording method for identification photography.

It may appear to many of those currently employed in the discipline of forensic identification that post-photography digital image processing appeared suddenly on the scene in the 1990s (or later), without past or provenance, and more or less concurrent with the digital revolution. Both technologies have required a steep and rapid learning curve, perhaps contributing to a level of discomfort in would-be users. Terms like "junk science" and "voodoo" have been used to describe the procedures by which net gains in image detail and clarity have been attained.

Nothing could be further from the truth. All digital image processing described in this book is based on quantifiable, reliable, repeatable, and transparent science. Some colleagues in the early stages of the science (from personal recollection and interaction) had displayed reluctance to conduct digital optimization actions, possibly because they didn't completely understand the science and didn't feel comfortable in explaining in court the processes by which evidence detail became clearer and more complete.

This chapter attempts to create a timeline of digital technology evolution, from the authors' perspective, the steps that led to the current acceptance and application of digital techniques to impression evidence, with the purpose of increasing the comfort level of practitioners, attorneys, and any others who encounter digital image processing. Another objective is to acknowledge the contribution of key contributors and pioneers known to the authors, those who have advanced the science through research, training, and pivotal casework, leading to landmark acceptance of this technology in courts of law. It does not purport to be a complete list, but it is one assembled from the experience and the perception of the authors.

Brian E. Dalrymple

I wish to make one point clear from the onset—that I am not a pioneer, but I had the very good fortune to encounter such individuals early in my career and profit from those associations. In 1977, I attended a conference of the Society of Photo-Optical Instrumentation Engineers (SPIE) in Reston, Virginia [1], to give the first international presentation on the use of lasers to detect evidence. Another speaker on the program was from the Itek Corporation [2], discussing computer analysis of the Zapruder film (Kennedy assassination). It was the first time I had been made aware of digital technology applied to forensic examination, and the effect was profound.

There were several significant challenges facing those who attempted to extract and optimize details from the Zapruder film. It was recorded on 8-mm movie film, of 1963 vintage, the negative size being only 4.5 × 3.3 mm. By comparison, the dimensions of a 35-mm negative (a relatively small format) was 36 × 24 mm. The camera was handheld, and the subject was moving. Lastly, the area of interest (JFK) was a very small part of the image area. When this area was enlarged, resolution became blurred, and fine detail disappeared. It resembled an impressionist painting, displaying the suggestion of something sinister rather than hard definable details and edges.

Viewing the digitally processed version for the first time had an indelible impression. In place of the blurred and obscured areas of color and tonal modulation, one could see crisp detail and much more defined color and tonal transitions. Simply put, this version revealed substantially more detail than the original film. If one can put aside for a moment that the subject of this film is the assassination of the president of the United States, it is a homicide investigation, and digital image processing is an outstanding tool. A question was asked during the presentation about the cost of the process. The response was, $10,000 per frame (in 1970s dollars). There are 16 frames per second in 8-mm movie film, and the presenter advised that 12 seconds of film were processed. It was clear that this technology was, for the present at least, well beyond the means of police investigators.

In the 1970s, the world experienced the beginning of a phenomenon that has been called the "pocket calculator syndrome." The first handheld portable calculators were specialized and expensive equipment, basic in function (add, subtract, multiply, and divide), costing hundreds of dollars. As the years passed, calculators became smaller, faster, more powerful, and substantially cheaper, to the point where they could be acquired for a small fraction of their initial cost. It was the writer's hope in 1977, that time, increased usage, and technological advancement would raise the quality and lower the price of the hardware and software, bringing it within the reach of forensic investigators. Today, it is apparent that this is exactly what has occurred. An extremely effective software/hardware package, based on Adobe Photoshop® and Media Cybernetics Image-Pro Premier, can be acquired for less than $10,000.

In 1987, I attended the International Forensic Symposium on Latent Prints, hosted by the Federal Bureau of Investigation (FBI) at Quantico, Virginia [3]. A most compelling presentation was given by E. R. German [4], U.S. Army CIL, Forest Park, Georgia. It was the first example known to me of achieving a significant increase in forensic image information through the application of computer processing, based on a packaged imaging software (Image-Pro, Media Cybernetics). Techniques including fast Fourier transform (FFT) and edge enhancement were discussed, supported by convincing before/after images to illustrate their value in revealing hidden or obstructed detail in latent fingerprints and other evidence images. Further, and critically, Image-Pro (the software system featured) was a package combined with hardware costing in total approximately $60,000 CAD. This was a huge reduction from the $2 million processing costs (in 1970s dollars) cited 10 years earlier at the Virginia conference in 1977.

Shortly thereafter, I was privileged to spend a week with Mr. German at his laboratory in Georgia and become acquainted with some of the tools used to optimize images. As a direct result of this visit, the first image processing system in a Canadian police agency was acquired in 1991, installed and configured by Pamela Ringer of Hunter GIS. The software, Image-Pro, featured many powerful algorithms that could optimize signal-to-noise ratio in evidence images. I and another colleague were selected for training in this new system. William Watling of Internal Revenue Service (IRS) Laboratory in Chicago generously agreed to provide training. All image capture at the time in our laboratory was analog (film or tube camera), and an analog/digital converter was an integral part of the hardware.

The system had been operational for less than six months when I was assigned a very high-profile case involving a police officer who had been charged with perjury and obstruction of justice. It had been alleged that a cigarette butt collected at a crime scene had been lost and replaced with another butt. The examination focused on comparing the subject cigarette butt with film photographs taken at the crime scene. These photographs were problematic in that the cigarette butt was a very small part of the area recorded, and when enlarged for comparison, the image was blurred and indistinct.

I used a sequence of procedures in completing this task, including line histogram, histogram equalization, and edge/transition enhancement. When the case was scheduled for court in 1991, it was learned that the defense would challenge the admissibility of the image processing techniques. This was the first time I would be required to tender evidence concerning computer image processing. Consistent with the policy for first-time fingerprint testimony in the Ontario Provincial Police, I forwarded my processing strategy, images, and conclusions to William Watling, who reviewed and approved my procedures and conclusions. Both he and Pamela Ringer were subpoenaed as witnesses.

Mr. Watling was called to the witness box first to explain and demonstrate computer processing to the court. He gave clear and powerful testimony (digital enhancement being unprecedented in Ontario courts, at minimum). The evidence I subsequently tendered, as to the optimization procedures I conducted and the opinions I formed, was uneventful and unconditionally accepted by the court and the defense without issue. To my knowledge, this was the first time in a Canadian court that such evidence had been tendered and accepted.

Edward Raymond German

As earlier stated, in 1991 the Ontario Provincial Police became the first police agency in Canada to acquire and use a computer image-processing system, all because of the fortuitous meeting and interaction with Ed German.

Throughout his career, German has demonstrated an intense and unremitting interest in all phases of identification science, including but not limited to evidence detection by laser, comparison of twin fingerprints, and, most relevant here, computer optimization of evidence images. He referred me to perhaps the first mention of digital image processing as it pertains to fingerprints. A review of Project Search was reported in *Identification News* in 1974 [5]. This project initially sought to explore the feasibility of computerizing criminal history in the field of identification, but it expanded to comprise other topics, including satellite transmission of fingerprints and an automated fingerprint search system involving Fourier transform and other mathematical techniques.

In November 1983 [6], German's article on analog/digital image processing appeared in *Identification News*. He described an affordable image processing system that was not beyond the resources of police agencies. He provided a list of terms, a digital lexicon unfamiliar to most fingerprint analysts of that era that decades later would be in everyday use. Techniques such as edge enhancement and density slicing illustrated the power of digital methods in optimizing images.

German offers a personal memory of the speed at which digital science was overtaking and altering the forensic world. He gave a presentation on image processing at a conference in 1984, citing the advantages in sensitivity of digital technology over human vision and film. He was challenged from the floor by experts from a federal crime lab who were extremely dismissive of digital techniques, labeling them as inferior to film. Three years later, the same individuals were seeking him out for training [3].

In 2016, German was presented with the Dondero Award, the highest honor awarded by the International Association for Identification. This honor is not granted on a yearly basis but only when and if it is deemed appropriate for the most significant and valuable contribution in the field of identification and allied sciences.

Robert D. Olsen, Senior

Robert Olsen, a native of Kansas, began a career in the military after high school, rising to the position of special agent in charge of the latent fingerprint division, with the U.S. Army Criminal Investigation Laboratory (USACIL) in Fort Gordon, Georgia. He was passionate and professional in the pursuit of cutting-edge fingerprint detection and the standards and training required to implement them effectively. He retired from the Army Crime Lab in 1978, returned to Burlingame, and continued his career with the Kansas Bureau of Investigation until his death on February 23, 1989.

Scott's Fingerprint Mechanics is arguably still one of the most important reference books on the detection, interpretation, and identification of latent fingerprints. An expanded, updated edition of the original text (published in 1951), it is impressively comprehensive not only in the range of subjects and techniques, but in how concisely and accurately they are reviewed and presented. Subject matter includes fingerprint detection on skin, laser-excited fingerprint luminescence, probability theory in terms of the likelihood of two individuals having the same fingerprint, and, most relevant to this book, optimization of fingerprints by computer techniques, including fast Fourier transform (FFT).

A case application of FFT in a San Diego homicide was described in an article published in 1972. A blood impression located on a bedsheet was obstructed by the weave pattern of the cloth, making it unsuitable for identification. The fingerprint image was sent to the Jet Propulsion Laboratory in Pasadena, California, where it was scrutinized by the Space Technology Applications Office and the Image Processing Laboratory. FFT, a cutting-edge technology, was used to suppress the weave pattern, affording a clearer view of the ridge detail. A chart was prepared, illustrating 17 points of comparison between the latent impression and the palm print of the accused.

The enhanced print was ruled inadmissible because, in the court's opinion, the People failed to establish that either the scientific principle or the method (FFT) was widely accepted by experts in the field of use. Furthermore, the court held that there was no scientific certainty about the results produced by the technique. It is significant that these finding related exclusively to the unprecedented use of the FFT method used by the Jet Propulsion Laboratory (JPL) and did not address the value of the enhanced impression. In fact, no latent fingerprint expert was permitted to testify regarding this impression. Fortunately, and partly due to the chemical enhancement of another area of bloody ridge detail on the bedsheet, a conviction in the case was ultimately secured.

Olsen expresses his own frustration with this finding and in doing so undoubtedly supports the sentiments of many others involved in the process. It is a perfect paradoxical "catch-22" situation, one of which Joseph Heller would be proud. A new technique cannot be used without general acceptance in the peer group, and it cannot gain acceptance without use. As Olsen states, under this rule, no new technique would ever gain acceptance, and there would be no progress [7].

The authors are not aware of any earlier case applications of FFT or other computer algorithms in the forensic identification discipline.

Pamela Ringer

In 1989, a new technique of image enhancement was introduced to an audience of fingerprint professionals at the Virginia State Fingerprint Examiner's Conference, an event that would prove crucial in the near future. One year later in Henrico County, the body of Dawn Bruce, a 22-year-old woman, was found in her apartment, sexually assaulted and stabbed to death. The only significant evidence located at the scene was a pillowcase bearing blood transfer. Closer scrutiny revealed ridge detail, lacking the clarity and continuity necessary for comparison to a known fingerprint. Later, at the Virginia Division of Forensic Science in Richmond, the pillowcase was processed with 1,8-diazafluoren-9-one (DFO), a cutting-edge, recently introduced protein reagent used to detect fingerprints on paper. The ridge detail was significantly improved, but the fabric weave pattern of the pillowcase still obstructed the impression, leaving it unidentifiable.

A photograph of the DFO-enhanced fingerprint was taken to Hunter GIS (Geographic Information Systems), where Pamela Ringer used specially developed computer software (including fast Fourier transform) to remove the weave pattern of the pillowcase, allowing fingerprint examiners for the first time to analyze clear and continuous ridge detail. In the modern world of compact, powerful, and fast computers, it is difficult to envision the challenges of using such software. Ms. Ringer spent approximately four hours in the enhancement process. The optimized fingerprint was compared to and identified as having been made by the left thumb of the suspect, Robert Knight.

Later in court, history was about to be written. Defense attorneys launched an aggressive attack on the scientific acceptance and reliability of the image processing software that was the key to identifying the impression. In a Frye hearing, Ms. Ringer explained and demonstrated the procedure to the court. Supporting her testimony was William Watling of the IRS Laboratory in Chicago, acknowledged at the time as one of the most experienced experts in the world in the field of image processing. The court ruled that the enhanced impression was indeed admissible and that the process had not altered the fingerprint pattern. This was the first instance in history of image processing techniques withstanding the challenge of a Frye hearing [8,9].

DNA evidence (another new science at the time) was later added to the powerful case against Knight, and he entered a guilty plea to the charges to avoid the death penalty and was sentenced to life in prison.

Ms. Ringer also attended court and tendered evidence in Canada in the previously mentioned perjury and obstruct justice case.

William J. Watling

After a term of active service with the U.S. Army, Bill Watling joined the Arizona Highway Patrol in 1969, and he quickly assumed duties in the classification and identification of fingerprints in the Department of Public Safety. He was promoted to supervisor of the Arizona DPS Latent Print Unit in 1973. He developed an interest in computers after reading about the JPL case previously mentioned. Over the course of the next 10 years, he experimented with noise removal in fingerprint images, in collaboration with both the JPL and Goodyear Aerospace, and enhanced many images for the San Diego Police Department [10].

In 1986, Bill moved to the IRS National Forensic Lab in Chicago, Illinois, and immediately began efforts to obtain image processing software from Hunter GIS, meeting Pamela Ringer in the process. The software was Image-Pro, which featured fast Fourier transform (FFT), a method for removing obstructive noise and pattern from images. He used the software to optimize images of fingerprints and handwriting in many criminal investigations, consulting, teaching, and tendering expert evidence for investigations in the United States, Canada, and Britain, including the Hayden case (see Erik Berg).

In 1989, he gave a presentation at the International Association for Identification (IAI) Conference, Pensacola, Florida, entitled "Where Is Forensic Digital Image Enhancement Today?"

His support and testimony concerning the image enhancement procedures in the Frye hearing, conducted for the 1990 murder investigation in Henrico County, were pivotal in the conviction of Robert Knight. In 1991, when the Ontario Provincial Police Forensic Identification Services initiated the first police computer image processing system in Canada, Watling was an excellent teacher, creating a firm foundation for a very successful and effective support service. As mentioned earlier, his testimony on digital enhancement in the original Canadian case was essential to the court's acceptance of digital evidence. He authored a paper in 1993, describing the FFT process and the details of the first Frye hearing acceptance of FFT procedure and evidence [11]. Watling took a position with the U.S. Border Patrol (Department of Homeland Security) in 2004, creating a latent print unit and conducting fingerprint image enhancements. He retired from that position in 2011.

As one of the first and most prominent practitioners in North America, Bill Watling has provided consultation and guidance for latent print examiners, prosecutors, and defense attorneys on issues of digital image enhancement. He has received many accolades for his work, not the least of which is a special commendation from the governor of the state of Arizona, and he holds distinguished member status with the International Association for Identification.

Erik Christian Berg

Erik Berg came into forensic science with a degree in graphic design. In 1984, he began a career in policing with the Pierce County Sheriff's Department (Washington), completing general law enforcement services, which included everything from criminal investigation and evidence collection to rescuing drowning victims as a member of the dive/rescue team. During this time, he displayed his innate computer abilities by writing a database program for tracking service and investigation calls and for the production of investigative reports.

In 1991, he moved to the Tacoma Police Department as a forensic specialist and had a full range of identification duties—crime scene investigation, evidence collection, fingerprint identification, and expert testimony. Once again, his gift for computer applications resulted in the creation of a software program used to digitally optimize and track images of latent fingerprints. The program included his version of fast Fourier transform, More Hits Pattern Removal Filter, an effective method for removing periodic noise (pattern) which occasionally obstructs fingerprint detail [12].

Erik Berg was one of the first forensic specialists to recognize, develop, and exploit digital technology in the optimization of images. He began with an imaging program called PhotoStyler (the Aldus Group) in the early 1990s, a competitor of Photoshop, converting when Adobe acquired Aldus. Berg was one of the first practitioners to understand the unique value and potential of FFT in removing pattern and periodic noise that routinely obstruct fingerprints and, dissatisfied with the existing version of FFT, began to develop his own.

This technique would prove to be crucial in resolving a murder investigation. In May 1995, a young woman named Dawn Fehring was found dead in her apartment. She had been sexually assaulted and strangled. One of the exhibits recovered from the crime scene was a bedsheet bearing blood transfer impressions. There appeared to be friction ridge detail recorded in blood, a fingerprint and a palm print. Amido black was used to further develop and darken the impressions, but due to the obstruction of the cloth weave, it remained unsuitable for comparison purposes.

Erik Berg, an authority on the use of computer image enhancement techniques, was consulted. He conducted digital photography of the exhibits and used his own software successfully to remove the obstructive weave pattern. The finger and palm impressions were subsequently compared to and identified as having been made by Eric Hayden, who was convicted of murder. The defense contested the digital processing of the impression, on the grounds that the technology was "novel" and had not gained general acceptance in the relevant scientific community. In 1998, the State of Washington Court of Appeals upheld the conviction and agreed with the prosecution that the technology used by Berg was not novel and is generally accepted by the relevant scientific community used to render an obstructed fingerprint identifiable, resulting in the conviction of the suspect [13]. The conviction was upheld on appeal, the first time in the United States for a case in which this technology played the pivotal role.

In 2001, Berg testified in another pivotal Frye hearing regarding the Victor Reyes case in Broward County, Florida, explaining the science and the strategy behind digital image enhancement. The court accepted his testimony, and the processes of optimization used in the case were accepted by the court. The prosecution was ultimately unsuccessful due to a chain of other issues, unrelated to the digital evidence [14].

David Witzke

David Witzke has more than 25 years of experience and excellence in the areas of Automated Fingerprint Identification System (AFIS) and forensic digital imaging. In 1994, as a training specialist with North American Morpho Systems (NAMSI), he met Erik Berg, who introduced him to the products of PC Pros, the More Hits software, in particular. Within two weeks, Witzke began employment with PC Pros as vice president of marketing. In 2003, PC Pros became Foray Technologies, and today Dave holds the position of vice president of program management [15].

During a high-profile homicide case (*State of Florida v. Victor Reyes*), impressions on duct tape, originally judged to be of no value, were optimized using PC Pros More Hits software and Adobe Photoshop. The defense aggressively challenged the admissibility of this procedure in a Frye hearing. The judge ruled that that digital enhancement of images is an accepted process throughout the forensic community. Dave Witzke was one of the witnesses testifying at this hearing [16].

For more than 20 years, Dave has lectured and conducted training workshops in digital image enhancement across the United States, including seven years at the FBI Training Academy in Quantico, Virginia, and more than six years at the British Columbia Institute of Technology. He also taught digital image processing for police agencies throughout the United Kingdom and in Switzerland. His name has become synonymous with excellence for training and consultation in digital image processing wherever in the world this technology is practiced.

E. Jill Smith

In 1999, after 10 year of service in the forensic identification Imaging department at York Regional Police (Ontario, Canada), I was ready to begin the transition from analog image capture and processing to digital image capture, storage and processing. My supervisor set up our first computer-imaging workstation, loaded with Adobe Photoshop, and handed me the Photoshop manual, advising me that in three weeks I was to attend the FBI Training Academy in Quantico, Virginia, to attend the forensic digital imaging of evidentiary photography course. At that time, I opened the manual and started to study, learning what a JPEG image was for the first time. Our service has come a long way since then.

Learning in Quantico was very exciting—the first in a long journey of forensic digital image training. The other participants in my Quantico class were impressed with the huge and powerful computer my police service had just purchased—an 18-gigabyte redundant array of independent disks (RAID) server with 128 megs of random-access memory (RAM) to serve images to investigators, covering all of York Region, Ontario (population approximately 1,000,000)! The first digital cameras boasted 5-megabyte capture in JPG and TIFF formats, and it quickly became apparent that our storage capability was woefully inadequate. Working with the IT department, the storage capability grew by leaps and bounds over the years. Current digital storage needs in York Region grow by almost 1 terabyte each month; with increased chip sizes, and increasing population and crime rates, that number will continue to increase.

David Witzke (Ski) was the first instructor to introduce me to the incredible world of forensic digital image processing at the FBI in Quantico, Virginia, in 1999. I have attended many variations of Ski's courses and workshops over the years, and I have enjoyed learning from him; I never walk away from an opportunity to hear Witzke speak or pick his brain on an imaging issue!

Early in my digital imaging journey, I was introduced to Brian Dalrymple, a practitioner in laser photography and digital image processing, and an ongoing dialogue and association began. Dalrymple introduced me to fast Fourier transform and Image-Pro software in his workshops and lectures. He conducted in-house training on more than one occasion, and I often consulted him as our unit grew and developed in digital technology very quickly. We came to routinely brainstorm ideas and image processing strategies, resulting in a teaching collaboration on digital workshops across North America.

Another significant influence in this field does not work in forensic science at all. Dan Margulis, originally a printing press expert, taught many advanced color correction courses, instructing on color mode blending and advanced sharpening techniques that I was surprised to find readily applicable to forensic digital imaging.

I would be remiss if I didn't also mention my supervisors. Rick Finn started this effort with the purchase of computers, software, and training. Because of him, we were among the first in Canada to become a fully operational digital unit. Dave Juck, my detective sergeant and friend for many years, continued to support our growth in digital technology, approving numerous courses and encouraging me to lecture and instruct as I grew in this field along with my unit. Currently, Brad Joice and John Jacobs continue with the progressive thinking that continues to push us forward in this technology. It is not possible for a service to be as successful as we have been in the implementation of digital technology without the ongoing support for members and their training, and for that I am grateful.

I am very excited about this book. My vision was and is to create a user-friendly *Workshop-in-a-Book* that can serve as an ongoing reference for any forensic investigator/photographer finding themselves in need of the ability to process files for analysis—a book and forensic image processing guide I wish I had at my disposal back when I began my journey in the digital world in 1999. I hope it serves you well.

Transition from Film to Digital Imaging

The last half century has seen unforeseeable, seismic change in the discipline of forensic identification, in every aspect of it. The 50 years before this displayed relatively modest advancement in detecting and recording fingerprints. There were improvements in fingerprint powder sensitivity and new capabilities in film and cameras, but from the 1970s forward, a host of fundamentally new technologies entered the arena, chemical, physical, and light applications that began to reveal more evidence on more types of surfaces than previously thought possible.

Until approximately the end of the twentieth century, film was absolute in its monopoly as the image recording technology for crime scenes and impression evidence. Cameras had also increased in complexity and function, with the introduction of such features as auto-exposure, auto-advance, and autofocus.

Concurrent with these major advances in evidence detection has been the conversion to a profoundly new science of evidence recording—the transition from film (analog) to the digital domain. The conversion has occurred at almost breathtaking speed. In less than two decades, film recording has been reduced from the indispensable and virtually exclusive means of recording crime scenes, fingerprints, shoeprints, and other evidence to something in the same class as daguerreotype, a pastime for hobbyists.

Certainly, these sweeping changes have transformed our society in all areas, but we now focus on the specific consequences of this transition for those who detect, record, and interpret impression evidence.

The goals of forensic professionals have not changed in a century. The forensic identification practitioner/photographer can be characterized as an expert in forensic signal recognition—the first, and perhaps the only one to detect and triage all and any physical evidence, at crime scenes or on exhibits, that may help to explain and solve a criminal case. It is not simply a question of perfect "20/20" vision. Seeing and noticing are different attributes. Two individuals can view an identical scene, receiving the same retinal image, but make significantly different observations and interpretations. The identification specialist must also recognize the presence of or the potential for other types of physical evidence, DNA for example, that fall outside his/her skill base.

Forensic professionals, through a combination of innate ability, training, and experience, are better positioned to detect and evaluate discreet evidence than a layperson. This involves recognition of the subtlest indication of a tread pattern or ridge detail and, through diagnosis and triage, identification of the objects and surfaces on which impression evidence might be found. From the moment of discovery, their goal is to optimize the signal-to-noise ratio, optically, chemically, and digitally. The quintessential example of a subject in which the signal-to-noise ratio requires no adjustment (beyond competent photography) is a fingerprint properly recorded in black ink on white paper.

From the moment a crime scene (or exhibit associated with it) is discovered until the evidence derived from it is tendered in court, a series of separate but interdependent steps is generated, through which impression evidence may be diagnosed, treated, photographed, and optimized. The steps include, but are not necessarily limited to:

- White, ultraviolet, laser, and forensic light source examination
- Physical treatment (powders, powder suspensions)
- Chemical methods
- Lighting techniques
- Photographic techniques (in the past, one could also include darkroom techniques), such as dodging and burning)
- Digital image processing

These steps are interdependent because the success of each is contingent on how well and how completely the previous ones were done. For example, regardless of the skill, ingenuity, and work ethic of the photographer, his or her efforts will be limited by shoddy, incomplete, inexperienced, or inept chemical processing. No individual, regardless of expertise, can use photography to undo or correct these shortcomings. One cannot draw five units of data from a four-unit container. The container had to contain five units in the first place. Such is the case with impression photography. Sadly, it is more than possible to draw only four units of data from a five-unit container.

Forensic science is correctly conservative in nature. Acceptance of new technologies happen at different rates within a discipline and depends on many factors. For example, ninhydrin was patented as a fingerprint development technique in 1955, but it did not become the exclusive frontline method for paper exhibits for more than a decade [17]. Excluding the forerunners and pioneers, it is reasonable to surmise that for most forensic identification practitioners prior to 2000, film photography represented the end of the road in terms of forensic reach. In other words, whatever you had managed to discover and record on film negatives and prints, in terms of impression evidence such as fingerprints and shoeprints, was pretty much as good as it was going to get. This was certainly the writer's experience in case assignments and in direct communication with police agencies of all sizes at the time.

With virtually all impression evidence now recorded digitally, however, there is considerably more opportunity for professionals to extend their detection reach beyond photography by further increasing signal-to-noise ratio in evidence images.

There are two strategies discussed in this book for exercising this extended reach:

- *Reactive*—To open a single image in Photoshop (or other image processing software) and apply noise reduction and contrast optimization techniques.
- *Proactive*—To identify and assess a challenging evidence photography scenario before taking the camera from its case or, in some cases, before choosing and applying development chemistry or techniques.

To capture multiple images of the subject, in the knowledge that combining such images will frequently offer better results than a single image, when they are combined in a program such as Photoshop.

Faced with this steep learning curve of technological change, forensic identification practitioners may experience the sensation of trying to ascend on the down escalator—they must be constantly moving to remain in the same place.

The Digital Edge

There are solid reasons for the meteoric rise of digital imaging and the subsequent demise of film. A darkroom, enlarger, and chemistry can be replaced with one camera, a computer with image-processing software, and a good printer. Processing and printing film, color film in particular, is costly and time-intensive on an ongoing basis. With a digital imaging system, there is no need for the chemistry and materials associated with film, much less the footprint previously occupied by the equipment required for processing and printing.

Spatial resolution of professional digital cameras currently equals (or exceeds) that of film, in 35 mm at least, which was a predominant format used in evidence photography [18]. Perhaps the more important question is—are the spatial and brightness resolution, dynamic range, and speed of digital cameras sufficient to record forensic subject detail faithfully and reliably? Judging by the virtually unanimous adoption of digital recording in forensic imaging, the answer is a resounding yes.

Certainly, there were darkroom techniques in the film era one could employ in atypical situations. Overexposure and underdevelopment was a technique for capturing an exceedingly high dynamic range subject in one exposure. One valuable technique for long exposures with film was taught at the Ontario Police College in the 1970s. The exercise was a night shoot with an extreme dynamic range—streetlights at one end of the dynamic scale and seemingly impenetrable shadows at the other. The camera was placed on a tripod and focused. The shutter was opened for an exposure of 30 minutes, ensuring that there would be density on the film in the shadow areas of the subject. Developing time was drastically reduced, enough for the shadow detail to develop but not enough for the highlight areas to threshold. This technique was called "expose for the shadows and develop for the highlights." Digital images captured in reduction applied with dead pixels removed (RAW) (14- to 16-bit) exhibit significantly higher dynamic range to begin with. Adjustments in RAW can reveal a surprising amount of detail in such high dynamic range (HDR) images (see Chapters 2 and 4).

Darkroom techniques, such as dodging and burning, were used to optimize clarity of fingerprint detail in the final print, but rarely if ever, in the writer's 32-year career in forensic photography, did they transform a tantalizing but unusable area of ridge detail into an impression suitable for identification.

One of the most valuable currencies in criminal investigation is time. Digital cameras offer immediate confirmation, while changes in lighting or exposure can still be easily made, that the image contains the desired information. In the film era, hours and sometimes days elapsed before the negatives were developed and printed.

The Identification Photographer

In the film era, the forensic photographer's duties were almost exclusively completed as soon as the exposures were taken and the film developed and printed. The data on the film consisted of irregularly shaped, silver halide crystals embedded in a thin gelatin layer [19]. They had either been exposed to light or not. Aside from choosing contrast grade in print paper and dodging/burning, there were few options in optimizing what was originally captured on the negative. In color photography, virtually all processing and printing was automated, completed within the police agency or by an external company. In contrast, a digital image can be compared to an iceberg, with most of its data remaining unseen in the spatial display.

How many shades of gray can be detected by the human eye? This question does not have a simple answer for several reasons. There is no agreement on the number, partly because it depends on viewing and lighting conditions, and partly because there is such a wide range of capability in human attributes. Gifted athletes can run a mile in four minutes, far beyond the ability of most people. Similarly, some sets of eyes have a high level of acuity and resolving power, while others depend on glasses or contact lenses to see clearly. A range of 30 gray intensities is a number that emerges frequently, although other estimates are considerably higher [20]. The diversity of opinion regarding color vision is much greater, ranging from 10 million upward, although some estimates are at 1 million [21,22].

An 8-bit imaging system can sense, display, and record 256 discrete grayscale intensities. If the image is recorded in red, green, blue (RGB), over 16 million different values can be detected. There several major differences between the digital camera and the human eye. Digital cameras quantify and store the intensity information received by the sensors in a complementary metal-oxide semiconductor (CMOS) chip. This data can be displayed whenever desired without change. In a 24-bit RGB image, more than 16.7 million different colors can be detected, stored, and displayed, each quantitatively different from the others.

Whatever the actual sensitivity to color or grayscale, there are monumental differences between the light sensors in a digital camera and the human eye. What we "see" is the result of the data received by the retinas of both eyes and processed by our brain. Our vision data is not storable or reproducible in the sense of exact values [20]. We do not have precise color or tonal memory. The brain cannot quantify or store visual data accurately. Our brain can be fooled by what it thinks it sees because so much of our vision is based on comparison and interpretation rather than objective values.

Figure 1.1 illustrates two lines, one horizontal and one vertical. Because of their placement, we perceive the vertical one to be longer, but they are the same length. In Figure 1.2, a gradient-filled circle features a horizontal bar of midtone gray. It is the same value of gray from one side to the other, but the left side appears darker because of the lighter background.

So much more image optimization is possible in the digital domain than could ever be accomplished with a film negative. The ability to exploit this fact relies on training and expertise that go beyond the camera. Identification photographers in some agencies use Photoshop or other software to process their own images. These duties are completed in other police departments by separate sections whose exclusive program mandate is the post-photography improvement of digital images.

The goal of this book is to aid both groups of professionals in this task by highlighting image diagnosis and triage with proven strategies and examples.

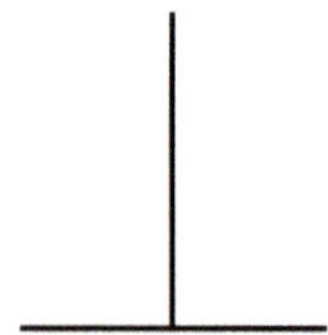

Figure 1.1 An example of the mind drawing an inaccurate conclusion of the respective lengths of two lines.

Figure 1.2 The horizontal bar is the same gray value from side to side. The brain uses comparative assessment rather than quantitative measurement to conclude that it is darker on the left side.

Finally, to underscore the validity of these techniques, selected examples of court acceptance are cited.

References

1. SPIE Technical Symposium East, Reston, VA, April 18–21, 1977.
2. F. Corbett, The Kennedy assassination film analysis, Itek Corporation, SPIE.
3. International Forensic Symposium on Latent Prints, FBI Academy Quantico, VA, July 7–10, 1987.
4. E. R. German, Computer image enhancement of latent prints and hard copy output devices. *Proceedings of the International Forensic Symposium on Latent Prints*, pp. 151–152, 1987.
5. COMSAT Proposed at Search Symposium, *Identification News*, pp. 3–5, July 1974.
6. E. R. German, Analog digital image processing of latent fingerprints. *Identification News*, pp. 8–11, November 1983.
7. R. D. Olsen, Sr., *Scott's Fingerprint Mechanics*, Charles C Thomas, Illinois, 1978a. Space age technology to the aid of the latent print examiner, *The Institute of Applied Science*, 1972.
8. P. Ringer, personal communication, 2016.
9. N. Tiller, The power of physical evidence: A capital murder case study, 1992, available at www.cbdiai.org/Articles/tiller_8-91.pdf
10. W. J. Watling, personal communication, 2016.
11. W. Watling, Using the FFT in forensic digital image enhancement, *Journal of Forensic Identification*, 43(6), pp. 573–583, 1993.
12. E. Berg, personal communication, 2016.
13. http://caselaw.findlaw.com/wa-court-of-appeals/1189825.html
14. *Florida v. Reyes* digital enhancement, available at http://www.forensictv.net/Downloads/legal/florida_v._reyes_digital_enhancement.pdf
15. D. Witzke, personal communication, 2016.
16. National Criminal Justice Reference Service, *Evidence Technology Magazine*, 1(2), pp. 16–19, July–August 2003.
17. *Journal of Criminal Law and Criminology*, 60(2), 1969, available at http://scholarlycommons.law.northwestern.edu/cgi/viewcontent.cgi?article=5596&context=jclc
18. Film vs. digital: A comparison of the advantages and disadvantages, May 26, 2015, available at https://petapixel.com/2015/05/26/film-vs-digital-a-comparison-of-the-advantages-and-disadvantages/
19. Camera lust: The Pentax K-01 and the Fuji X-Pro1, available at http://www.microcosmologist.com/blog/camera-lust-the-pentax-k-01-and-the-fuji-x-pro1/
20. Cameras vs. the human eye, Cambridge in colour, available at www.cambridgein colour.com/tutorials/cameras-vs-human-eye.htm
21. Humans can only distinguish between about 30 shades of gray, February 19, 2015, available at http://www.popsci.com/humans-can-only-distinguish-between-about-30-shades-gray
22. Number of colors distinguishable by the human eye, available at https://hypertextbook.com/facts/2006/JenniferLeong.shtml

Establishing Integrity of Digital Images for Court

2

The digital revolution is far more significant than the invention of writing or even of printing.

Douglas Engelbart

There has been much discussion in recent years with regard to image integrity and authentication as it applies to digital imaging, sparked by the massive movement in the forensic discipline over the last two decades from photography in the analog (film) domain to the digital domain. This chapter will touch on the supporting bodies that have guided the forensic community to make that transition seamlessly and successfully, the rules of evidence that dictate workflows, and how we are able to ensure and maintain the integrity of images used for investigation, analysis, and court.

As a forensic community, we approached the transition from analog film to digital imaging cautiously, and rightly so; we are all aware that digital images can be easily and almost seamlessly edited and manipulated. Hollywood dramatizations, magazine ads, and television commercials are increasingly sophisticated in their digital artistry and are good examples of why we must be diligent in our business to maintain the continuity and the integrity of the images we capture and process for forensic purposes.

That said, there is no need to be apologetic or afraid of this sound technology. The technology used to process digital images for analysis today was developed by the National Jet Propulsion Lab in the early 1960s (for a detailed summary of the history of digital image processing, refer to Chapter 1), allowing scientists to study data from the images captured of outer space. Weather research and forecasting scientists utilize digital imaging technology to predict and follow weather patterns, and in the medical field, we stake our lives every day on the digital imaging technology available to x-ray, scan, and monitor our bodies (CAT scans, radiology, ultrasounds, etc.). Forensic digital imaging technology has survived numerous Frye hearings, and there is a plethora of case law supporting this science, some of which will be outlined in Chapter 8.

The support of many groups has enabled forensic services internationally to modernize their workflows effectively, and while digital technologies have introduced many challenges that must be overcome (such as storage of large amounts of data, backup systems, training of staff, and writing of new policies), it also came with many benefits—some of which we had not anticipated at the outset;

for in addition to the traditional types of adjustments such as brightness and contrast, we discovered that digital image processing introduced new processing techniques such as channel blending, image subtraction, high dynamic range (HDR) merge, and fast Fourier transform (FFT) that often result in image detail being clarified beyond what was expected or done before. These new tools not only potentially increase the success of investigators in making an identification of a suspect in a crime but, just as importantly, if not more, in eliminating persons of interest early on in an investigation, saving investigators' time, pointing them in a more accurate direction, and relieving innocent individuals of undue stress.

There are many references and supporting bodies to aid forensic investigators and organizations in finding industry standards, general guidelines, and recommendations for many, if not all, aspects of this discipline.

Scientific Working Group on Imaging Technology (SWGIT)

SWGIT was formed by the Federal Bureau of Investigation (FBI) in 1997 [14]. This group included many professionals in the field of imaging from federal, state, and local law enforcement agencies, the American military, academia, international law enforcement agencies, and other researchers. Their mission was to help facilitate integration of digital imaging technology by providing best practices and guidelines for capture, storage, processing, analysis, transmission, and archiving of forensic digital images.

Overview of SWGIT and the Use of Imaging Technology in the Criminal Justice System, Section 1.4 *Admissibility of Digital Images.*

Digital imaging is an accepted practice in forensic science, law enforcement, and the courts. Relevant, properly authenticated digital images that accurately portray a scene or object are admissible in court. Digital images that have been enhanced are admissible when the enhancement can be explained by qualified personnel [1].

Over almost two decades, more than 20 documents have been created to guide forensic practitioners with best practices and guidelines for creating standard operating procedures with respect to digital imaging, image processing practices, and recording of processing steps made to the image, just to name a few. The work of this esteemed group over the years has had a direct impact on the transition from traditional film technology to digital photography and has further allowed us to learn and grow in this new technology so that we can produce better, clearer, and more complete images while maintaining their integrity.

Although SWGIT ceased operations due to funding in May 2015, its documents remain relevant and may still be accessed at: https://www.swgit.org/documents. The Scientific Working Group on Digital Evidence (SWGDE) [2] created a space for the SWGIT work and has added three subcommittees (photography, image analysis, and video analysis). These subcommittees are filled with many former SWGIT members. SWGDE is working to keep current with the issues concerned with digital imaging. It has revised and posted many former SWGIT documents; these SWGDE digital imaging documents can be accessed at: https://www.swgde.org [2].

Also working in the same digital imaging space is the Organization of Scientific Area Committees (OSAC) [3] Subcommittee on Video/Imaging Technology and Analysis (VITAL). OSAC VITAL's mission is to create more rigorous consensus standards from the best practices developed by the SWGs and other groups. This includes updates and some replacements to the SWGIT documents over the coming years.

Organization of Scientific Area Committees (OSAC)

OSAC [3] was formed in February 2014 as a part of an initiative by the National Institute of Standards and Technology (NIST) and the Department of Justice (DOJ) to strengthen forensic science in the United States. OSAC's mission includes supporting the development and publication of consensus driven standards and guidelines, which will be posted on the OSAC Registry of Approved Standards and the OSAC Registry of Approved Guidelines. Each of the scientific area committees (SACs) is comprised of forensic practitioners, academic researchers, measurement scientists, and statisticians, allowing participation, contribution, and comments from all stakeholders. The OSAC's aim is to identify and promote technically sound, consensus-based, fit-for-purpose documentary standards that are based on sound scientific principles.

Standards or Guidelines—What's the Difference?

OSAC definition of standard:

> *Standards specify uniform methods, actions, practices, processes, or protocols. Compliance is mandatory and modified only under unusual circumstances.*

OSAC definition of guideline:

> *... strongly recommend methods, actions, practices, or processes to consider in absence of applicable standards or best practices that are not mandatory.* [4]

There Is a Standard on Writing Standards!

OSAC uses the American National Standards Institute ANSI/SES 1:2013, Recommended Practice for the Designation and Organization of Standards as a guide for drafting standards, which outlines how to create standards properly, so they meet the professional standard required to give documents credibility in their respective fields.

Existing standards and guidelines (such as the SWGIT documents) are subjected to the OSAC Registry Approval Process of Published Standards and Guidelines and are measured against the following criteria: analysis of technical merit, openness of the development process, consensus, harmonization, and impact on the forensic science community. Multiple levels of OSAC participate in the evaluation of these documents against these criteria. If all levels agree that these criteria are met, a public notice of intent is published, including an open comment period to solicit feedback from all stakeholders. All of this is reviewed and considered before voting whether or not to include the standard or guideline on the registry.

Many existing standards and guidelines of the SWG groups have not yet been approved for the OSAC registry; OSAC stresses that this does not mean that they are invalidating their use, and further, that the current SWG documents should be used until they are superseded by OSAC. Many of these existing SWG and other best practice documents have already been posted on the OSAC websites for each of the appropriate disciplines as "Discipline-Specific Baseline Documents"; they are available there as reference materials that best reflect the current state of the practice within their respective disciplines. OSAC also has a Catalog of External Standards and Guidelines, which is a collection of standards, guidelines, and other documents applicable to forensic science. Additionally, OSAC is reviewing these standards for inclusion of the registry and has stated that these standards and guidelines may not have been recommended yet, may be in the process of being validated, or may not have met all of the necessary criteria to be approved for inclusion on the registry and is undergoing an editing process with the appropriate SAC. OSAC members may propose the modification of an existing nonapproved standard or the creation of a new one.

How Does OSAC Differ from the SWG Groups?

For a standard to reach the OSAC registry, it must survive a much more rigorous process. The standard is vetted by the OSAC legal, human factors, and quality resource committees, statisticians, and other OSAC subcommittees that are impacted, as well as be processed through a standards development organization that includes a public comment period. These resources were not part of the SWGs structure. "*SWG Standards were often informal (and were actually guidelines), not defined in a way that would meet the requirements of a standard from an SDO (Standards Development Organization)*" [5].

The OSAC consists of five scientific area committees (SACs), which report to the Forensic Science Standards Board (FSSB):

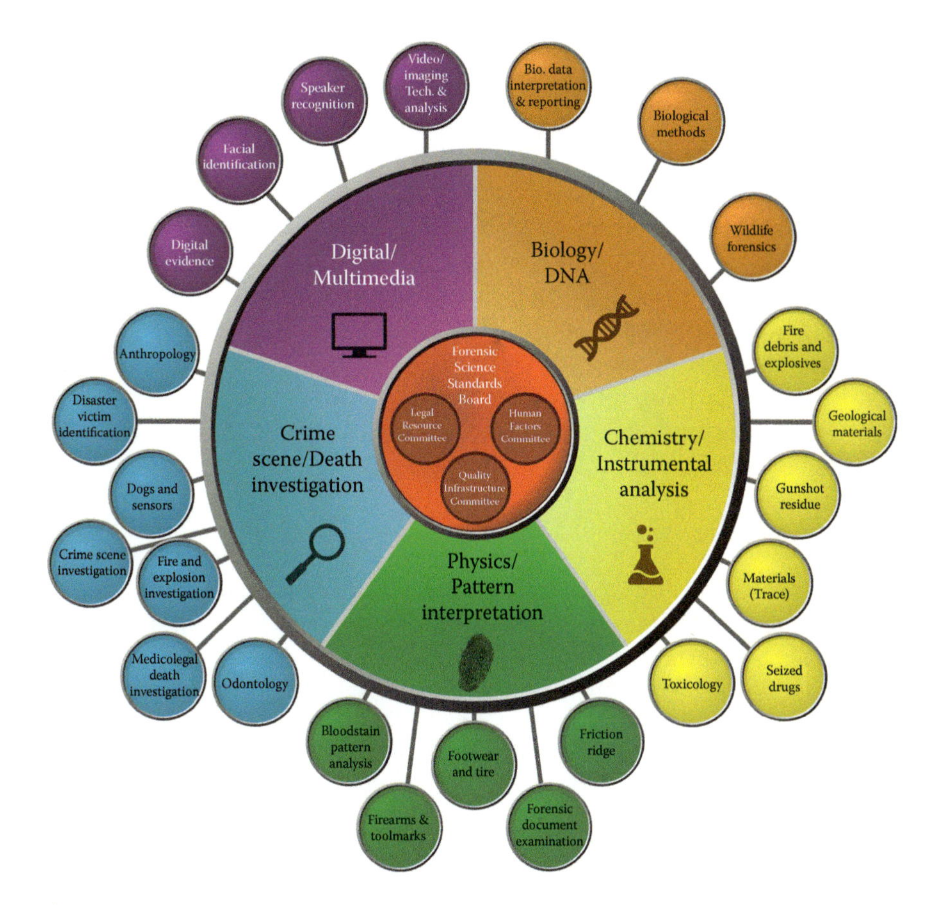

Figure 2.1 OSAC and SAC organizational chart.

1. Biology/DNA Committee
2. Chemistry/Instrumental Analysis Committee
3. Crime Scene/Death Investigation Committee
4. Digital/Multimedia Committee
5. Physics/Pattern Interpretation Committee

Figure 2.1 illustrates the structure of the OSAC. Each SAC oversees several subcommittees within their specialty. Please refer to NIST's website, and sign up to receive news about NIST Forensic Science and OSAC and/or download an application to become a member or affiliate of one of the SACs: http://www.nist.gov/forensics/osac/index.cfm.

Scientific Working Group on Digital Evidence (SWGDE)

Although SWGIT ceased operations due to funding, SWGDE is still funded and very active, working in collaboration with the digital multimedia SAC (of OSAC) to develop guidance documents on the subdisciplines of digital evidence. SWGDE, like most SWGs, can address current and emerging issues affecting the forensic science community more quickly due to its structure, which doesn't include all the actions an OSAC standard is required to go through. SWGDE is divided into six working committees, comprised of individuals actively engaged in the field of digital and multimedia evidence:

1. Audio Committee
2. Forensic Committee
3. Imaging Committee
4. Photography Committee
5. Quality Committee
6. Video Committee

The mission of SWDE:

The Scientific Working Group on Digital Evidence brings together organizations actively engaged in the field of digital and multimedia evidence to foster communication and cooperation as well as ensuring quality and consistency within the forensic community [2].

SWGDE documents may be accessed at: https://www.swgde.org/documents.

Rules of Evidence in Both the United States and Canada

(As related to digital images.)

The use of digital images in court as evidence is determined by rules of evidence and case law. For both areas, digital images have been admissible in court since at least 1991, with guidance groups such as the Scientific Working Group for Imaging Technology supporting the transition from analog to digital, providing the guidance needed to allow law enforcement agencies to do so successfully. The requirements for admitting digital images into court are the same as they were with traditional photography.

Federal Rules of Evidence for the United States

In the United States, courts follow the Federal Rules of Evidence, and while most states have their own Rules of Evidence, many have adopted the wording of the Federal Rules. Rules 1001–1008 deal with definitions pertaining to originals and duplicates, the requirements of originals, exceptions that allow for copies of originals, and testimony of individuals authenticating the contents of a photograph in court.

Rule 1001. Definitions That Apply to This Article
In this article:

1. *A "writing" consists of letters, words, numbers, or their equivalent set down in any form.*
2. *A "recording" consists of letters, words, numbers, or their equivalent recorded in any manner.*
3. *A "photograph" means a photographic image or its equivalent stored in any form.*
4. *An "original" of a writing or recording means the writing or recording itself or any counterpart intended to have the same effect by the person who executed or issued it. For electronically stored information, "original" means any printout—or other output readable by sight—if it accurately reflects the information. An "original" of a photograph includes the negative or a print from it.*
5. *A "duplicate" means a counterpart produced by a mechanical, photographic, chemical, electronic, or other equivalent process or technique that accurately reproduces the original.*

(Pub. L. 93–595, § 1, Jan. 2, 1975, 88 Stat. 1945; Apr. 26, 2011, eff. Dec. 1, 2011.)

Rule 1002. Requirement of the Original
An original writing, recording, or photograph is required in order to prove its content unless these rules or a federal statute provides otherwise.
(Pub. L. 93–595, § 1, Jan. 2, 1975, 88 Stat. 1946; Apr. 26, 2011, eff. Dec. 1, 2011.)

Rule 1003. Admissibility of Duplicates
A duplicate is admissible to the same extent as the original unless a genuine question is raised about the original's authenticity or the circumstances make it unfair to admit the duplicate.
(Pub. L. 93–595, § 1, Jan. 2, 1975, 88 Stat. 1946; Apr. 26, 2011, eff. Dec. 1, 2011.)

Rule 1004. Admissibility of Other Evidence of Content
An original is not required, and other evidence of the content of a writing, recording, or photograph is admissible if:

1. *All the originals are lost or destroyed, and not by the proponent acting in bad faith;*
2. *An original cannot be obtained by any available judicial process;*
3. *The party against whom the original would be offered had control of the original; was at that time put on notice, by pleadings or otherwise, that the original would be a subject of proof at the trial or hearing; and fails to produce it at the trial or hearing; or (d) the*

writing, recording, or photograph is not closely related to a controlling issue.

(Pub. L. 93–595, § 1, Jan. 2, 1975, 88 Stat. 1946; Mar. 2, 1987, eff. Oct. 1, 1987; Apr. 26, 2011, eff. Dec. 1, 2011.)

Rule 1005. Copies of Public Records to Prove Content

The proponent may use a copy to prove the content of an official record—or of a document that was recorded or filed in a public office as authorized by law—if these conditions are met: the record or document is otherwise admissible; and the copy is certified as correct in accordance with Rule 902(4) or is testified to be correct by a witness who has compared it with the original. If no such copy can be obtained by reasonable diligence, then the proponent may use other evidence to prove the content.

(Pub. L. 93–595, § 1, Jan. 2, 1975, 88 Stat. 1946; Apr. 26, 2011, eff. Dec. 1, 2011.)

Rule 1006. Summaries to Prove Content

The proponent may use a summary, chart, or calculation to prove the content of voluminous writings, recordings, or photographs that cannot be conveniently examined in court. The proponent must make the originals or duplicates available for examination or copying, or both, by other parties at a reasonable time and place. And the court may order the proponent to produce them in court.

(Pub. L. 93–595, § 1, Jan. 2, 1975, 88 Stat. 1946; Apr. 26, 2011, eff. Dec. 1, 2011.)

Rule 1007. Testimony or Statement of a Party to Prove Content

The proponent may prove the content of a writing, recording, or photograph by the testimony, deposition, or written statement of the party against whom the evidence is offered. The proponent need not account for the original.

(Pub. L. 93–595, § 1, Jan. 2, 1975, 88 Stat. 1947; Mar. 2, 1987, eff. Oct. 1, 1987; Apr. 26, 2011, eff. Dec. 1, 2011.)

Rule 1008. Functions of the Court and Jury

Ordinarily, the court determines whether the proponent has fulfilled the factual conditions for admitting other evidence of the content of a writing, recording, or photograph under Rule 1004 or 1005. But in a jury trial, the jury determines—in accordance with Rule 104(b)—any issue about whether:

1. *An asserted writing, recording, or photograph ever existed;*
2. *Another one produced at the trial or hearing is the original;*
3. *Other evidence of content accurately reflects the content.*

(Pub. L. 93–595, § 1, Jan. 2, 1975, 88 Stat. 1947; Apr. 26, 2011, eff. Dec. 1, 2011.) [6]

Canada Evidence Act

Authentication

Section 31.1 *Any person seeking to admit an electronic document as evidence has the burden of proving its authenticity by evidence capable of supporting a finding that the electronic document is that which it is purported to be.* 2000, c. 5, s. 56 [6]

Authentication of a photographic image is often achieved with the testimony of the photographer or a witness who can testify that the image is an accurate representation of the evidence or scene it is claimed to be.

Best Evidence

Section 31.2 *(1) The best evidence rule in respect of an electronic document is satisfied*

1. *On proof of the integrity of the electronic documents system by or in which the electronic document was recorded or stored;*
2. *If an evidentiary presumption established under* section 31.4 *applies* [7].

Weighing Evidence—Appendix A: The Rules of Evidence and the Canada Evidence Act

A.3. The Best Evidence Rule

A.3.1 The Rule:

"The law does not permit a man to give evidence which from its very nature shows that there is better evidence within his reach, which he does not produce" [8].

A.3.2. Application of the Rule

While this rule originally applied to all evidence, it has been restricted in its application to documentary evidence: if the original document is available, it must be produced. Otherwise, all relevant evidence is admitted into evidence, and whether it is the best evidence available, simply goes to weight [8].

The Best Evidence Rule in Canadian law requires the proponent of evidence to produce the best evidence available to that party. The spirit of this dictum states that we have an obligation to provide the court with the clearest, truest, and most complete information possible. If we *can* process a fingerprint or footwear impression in order to make it easier to visualize, then under the Best Evidence Rule, we are obligated to do so.

The history of the Best Evidence Rule can be traced back as far as the eighteenth century. In English court, *Omychund v. Barker* (1745), Lord Harwicke stated that no evidence was admissible unless it was the *"the best that the nature of the case will allow"* [9].

Presumption of Integrity

Section 31.3 *For the purposes of subsection 31.2(1), in the absence of evidence to the contrary, the integrity of an electronic documents system by or in which an electronic document is recorded or stored is proven*

1. *By evidence capable of supporting a finding that at all material times the computer system or other similar device used by the electronic documents system was operating properly or, if it was not, the fact of its not operating properly did not affect the integrity of the electronic document and there are no other reasonable grounds to doubt the integrity of the electronic documents system;*
2. *If it is established that the electronic document was recorded or stored by a party who is adverse in interest to the party seeking to introduce it;*
3. *If it is established that the electronic document was recorded or stored in the usual and ordinary course of business by a person who is not a party and who did not record or store it under the control of the party seeking to introduce it.*

2000, c. 5, s. 56 [7]

Image Integrity

Dictionary definition of integrity
1 uprightness; honesty. 2 wholeness. 3 the condition of being uncorrupted [10].

While images presented in court may be authenticated by the photographer or a witness testifying that any given image is an accurate representation of what it is claimed to be, image integrity addresses the policy, workflow practices, and security steps taken by the forensic photographer and/or forensic examiner responsible for photographing evidence and crime scenes, ensuring that images are protected, preserved, and unchanged in their whole entirety from the moment the shutter button is released, until its final disposition. When an image is processed for image analysis, the result must be a whole new file identified as a processed version of the original, and the integrity of the new file must also be protected.

There are many means of establishing integrity, and a law agency may use several methods to do so; the list that follows is by no means exhaustive. The methods your agency chooses depends on many factors, such as the size of the law enforcement agency, the level of training of staff, cooperation and involvement of the information technology department, and budget restraints.

Methods for Maintaining Integrity

1. *Standard operating procedures*—It is recommended that all forensic photographers and/or forensic examiners tasked with photographing scenes and/or evidence, draft a standard operating procedures manual (SOP) outlining the steps to be taken to maintain security and chain of custody of images. All versions of an organization's operating procedures should be archived for reference purposes. As important as it is to draft such procedures, it is equally important to ensure that members are trained in the content of the procedures and that copies are made readily available for reference.
2. *Physical security*—Physical barriers to bar unauthorized access to image data may include locked doors, security guards, and other physical restrictions. Additionally, a good plan for the storage needs of any forensic organization should take environmental factors into account. Humidity, temperature, dust/dirt, electrical surges, fire, floods, and electromagnetic field interference are all risks that should be considered when creating a long-term storage plan.
3. *Redundancy and backup*—While redundant backups of media should be stored in an alternate location to prevent loss in case of natural disasters and other environmental events, the more likely scenario is loss due to a system failure. As a point in case, in 2004, a Canadian law enforcement agency lost a month's worth of imaging data due to a relatively simple error. The information technology (IT) department was happy to refer to their backup tapes to recover the lost data; however, it was at that point they discovered that all of the backed-up data was corrupt and unrecoverable. Thankfully, the forensic unit had, as per their policy, created an independent, internal office backup on CD (separate location, stored in locked cabinet), and although it was a somewhat time-consuming endeavor, they were able to recover the entire month of data. Needless to say, it was a very important learning experience that led to some minor modifications to operating procedures, and a closer working relationship with the IT team.
4. *Computer security*—Agency security policies should address issues such as digital electronic access tracking, member password access, restricted access, virus protection, and protective firewalls, often a task largely assumed by the IT department.
5. *Retaining older versions of hardware and software*—An organization would do well to consider retaining original hardware and software used to create and access digital data to ensure accessibility. Imagine not having the ability to access processed images and/or records of their notes from an older case because they are not compatible with new systems and hardware!

Storage of Digital Data

The archiving of digital image captures and processed images is required to ensure that the stored data is accessible for future use. One consideration for choosing a long-term storage method is the length of time the data must be stored to satisfy statutory requirements and agency policies.

1. *Compact flash (CF) cards, security digital (SD) cards, and flash USB sticks*—CF and SD cards are common storage media for our capture devices and are considered highly reliable due to a lack of moving parts. An incredible amount of data may be saved on these and USB flash memory devices, but they are not suitable for mass storage of files in the long term.
2. *Optical storage (CD/DVD)*—Write-once optical media is recommended along with burning multiple copies (redundancy backup). Used alone, this method of storing digital data is considered to be inadequate for long-term archiving. The lifespan of optical media begins at time of manufacture, and there are concerns regarding the longevity of this medium; environmental factors such as humidity and temperature may be factors. This option is becoming less viable every day, as discs are quickly becoming obsolete.
3. *Magnetic storage*—Some types of magnetic tape have been shown to be reliable in the long-term storage of digital data; consideration should be given to magnetic tape formats designed specifically for long-term archiving. Redundant array of independent disks (RAID) combines multiple disk drive components for the purposes of "data redundancy" or backup reliability.
4. *The Cloud*—While we are not ready to move into this area of storage capability because of the obvious security risks involved, one must wonder how this storage medium may evolve over the years to possibly make it possible. Time will tell.

It is recommended that an agency work closely with its IT team to discuss archiving requirements and issues and build an archival structure that best suits its needs. A hybrid of more than one storage option may be adopted in the interest of redundancy. Any archiving plan developed with your IT team may include compression, but care must be taken to ensure that compression is *lossless*, and that systems are able to decompress the data for access at a later date.

Image Authentication

Not to be confused with image integrity, image authentication is the ability establish that an image is a reasonable and accurate representation of what it is claimed to be. It is common practice in court to call on the photographer to enter the photographic evidence, testifying that the images are an accurate representation of the scene. It is also possible that witnesses may testify to the accuracy of the images submitted during the course of their testimony.

SWGIT definition of "forensic image authentication":

> *Forensic image authentication is the application of image science and domain expertise to discern if a questioned image or video is an accurate representation of the original data by some defined criteria ...*
> *... Questions involved in authentication include issues of image manipulation, image creation, and consistency with prior knowledge about the circumstances depicted.* [11]

Without either the photographer or a witness to testify to the validity of an image, some experts may (or may not) be able to determine the authenticity of an image by analyzing the metadata, detecting manipulation by evaluating textures, shadows, highlights and size relationships, searching for image creation clues such as unrealistic or unnatural image features, and/or looking for inconsistencies that may flag a continuity issue, such as missing images in sequence, or concerns that the image data was not preserved as per policy to ensure its integrity.

Without the photographer or a witness, determining the authenticity of a digital image may be somewhat subjective and difficult, if not impossible to prove mathematically. Detecting staging and/or manipulation may be a series of clues that point to possibilities or probabilities.

File Formats

There are a few file formats that are necessary for every forensic crime scene investigator/photographer to be aware of:

1. *JPEG (joint photographers expert group)*—One of the most commonly used imaging file formats, JPEG is very versatile. Because it can be compressed to a smaller size, it is faster to download and easier to store, making it a popular choice for Web imaging software, inserting into Word documents, electronic presentations, and email. It is compatible with almost everything that you can think of where digital images are concerned. There are a few important caveats, however, and as a forensic professional, it is important to be aware of them:
    a. Processing: In-camera processing is applied automatically when you press the shutter release. This means that white balance, sharpening, saturation adjustments, brightness, and contrast are

automatically applied; these are *permanent* adjustments made to the file that cannot be undone.

b. Compression: *All* JPEGs are compressed to a smaller file using sophisticated algorithms. The compression level may be controlled in the camera settings anywhere from high compression (low image quality), to low compression (high image quality), and because of this process, some of the original image data is permanently lost—this is called *lossy* compression. As ominous as that sounds, most members of society are very familiar with JPEGs, using them every day; they understand that JPEG compression does *not* change apples to oranges. What *we* need to understand, however, is that JPEG compression may cause noise and distracting artifacts, particularly around edge details within an image file, and the higher the compression rate, the worse it gets. SWGIT distinguishes between Category 1 images used for documentary purposes and Category 2 images used for scientific analysis purposes (best practices for documenting image enhancement [12]). Category 1 images may be captured in JPEG format, and this is an acceptable practice among many law agency services that must consider storage space and ease of sharing data with the courts. Category 2 refers to images intended for analysis, requiring lossless image capture (RAW). Most digital single-lens reflex cameras have multiple image compression settings allowing the photographer to control the level of compression and often will allow for JPEG plus RAW capture at the same time.

2. *TIFF (tagged image file format)*—TIFF is another commonly used *lossless* image file format, which makes it an excellent choice for capturing (scans) and saving Category 2 images intended for investigative and scientific analysis. These files are very large and require more storage space than JPEG or even RAW; they will support up to 4 gigabytes of data. While compatible with most imaging programs, TIFFs aren't as compatible with programs such as Microsoft Word and PowerPoint, and their large size makes them slow to open or transfer. In Photoshop, they support adjustment layers, 16-bits-per-channel, and high dynamic range, 32-bits-per-channel images. When saving images in the TIFF format, you have the option to choose LZW, ZIP, and JPEG compression methods or no compression at all.

   a. LZW (*Lempel-Ziv-Welch*) is a lossless compression method; the lossless compression algorithm allows a mathematically exact reconstruction of an image's pixel values. It works most effectively on large areas of a single color and tone.

   b. ZIP is also a lossless compression method; be aware that not all software programs can recognize these compressed files. As with

LZW, ZIP compresses most effectively with large areas of a single tone and color.

c. JPEG, again, is a *lossy* compression method and is not recommended for any images intended for analysis (Category 2 images).

3. *RAW*—RAW is not an acronym but simply means "unprocessed raw image data." This file format is lossless and is considered to be a digital negative. It is a read-only file and cannot be overwritten, but rather, it may be saved as something else (JPEG or TIFF, for example) in addition to the original file. It is not compatible with all programs that work with digital images. Even Windows is unable to interpret and display a RAW thumbnail in the browser. A RAW image file converter is required to view and process your image before you can save it in a format that can be used in other software systems. You may use either the proprietary converter included with the purchase of the camera or an imaging program such as Adobe Photoshop that has a RAW file converter built in (Adobe Camera Raw). While that may seem to be a great disadvantage, consider the power of control at your fingertips when you have a RAW file to work with. RAW supports color bit-depth up to 16 bits, meaning that there are many more tonal and color values to work with, allowing for much greater control at the beginning of the processing stage. In Adobe Camera Raw (ACR) or another RAW file converter, you can adjust white balance and increase or decrease exposure. Shadows may be lightened, and highlights may be darkened. Contrast can be increased or decreased. All these adjustments are handily stored in an XMP sidecar file that is saved alongside your original RAW. Always save and keep these two files together with your casework, particularly if you are processing an image intended for analysis. It is the first recorded step of your image optimization process.
4. *PSD (Photoshop document)*—This is the only format besides large document format (PSB) that fully supports all Photoshop functions and file sizes up to 2 gigabytes. Choosing the *maximize file compatibility* option when saving a PSD file prompts the software to create a composite version of the layered file, allowing it to be read by other applications and enabling a faster load time. Will support adjustment layers, 16-bits-per-channel, and high dynamic range 32-bits-per-channel images [15].

Image Processing—Tracking Methods

SWGIT Guidelines for Image Processing, Section 5—It is the position of the Scientific Working Group on Imaging Technology (SWGIT) that any changes to an image made through image processing are acceptable in forensic applications provided the following criteria are met:

1. *The original image is preserved.*
2. *Processing steps are documented when appropriate, in a manner sufficient to permit a comparably trained person to understand the steps taken, the techniques used, and to extract comparable information from the image.*
3. *The end result is presented as a processed or working copy of the image* [13].

Definitions:

Primary image—*The first instance in which an image is recorded onto any media.*
Original image—*An accurate and complete replica of the primary image, irrespective of media.*

Once the digital primary images are established and properly archived, the issue of image processing, reproducibility, and integrity should be addressed. Several options exist for documenting and securing evidence images and the processing procedures applied to them.

Ready to Digitally Process an Image?

The first step is to be working on a *copy* of the original RAW image in Adobe Camera Raw or the proprietary software recommended for the RAW images captured with your camera brand. The RAW files may be adjusted for contrast, exposure, color saturation, highlight adjustment, shadow adjustment, clarity, and white balance, just to name a few that are most helpful at the first stage of image processing. Making adjustments in ACR results in a sidecar file being stored alongside the original file, called an "extensible metadata platform" or, XMP file. This file stores all adjustments made there, and if the RAW file is re-opened, those settings will be re-applied to the original; however, remember that the file itself cannot be overwritten; it is giving you a preview of the original with the ACR adjustments applied, and you will have to save these results as something else (TIFF, JPEG). Case files should include the original RAW, the XMP file, and a lossless file such as TIFF resulting from the adjustments made manually to the RAW file.

The second step of image processing is often calibration of the image to size it as an accurate life-size model to enable search options in the Automated Fingerprint Identification System (AFIS). The image should then be saved as a calibrated version of the original. At this point, you are prepared to begin processing the image for scientific analysis and must choose a method of recording the steps taken to make it easier to visualize.

Methods of Tracking Processing Steps

Metadata

What is metadata? Metadata is any data that helps describe the content or characteristics of a file. This includes everything from the basic information of a file such as file format, pixel dimensions, width, height, and date taken, GPS coordinates, photographer and copyright notices, camera settings, and even, potentially, any adjustments and changes made to the file within Adobe Photoshop. The processing details stored here depends on the settings in the Edit > Preferences > History Log of Photoshop (more on that next).

EXIF (exchangeable image file format) and IPTC (International Press Telecommunications Council) are the most common header formats for storing metadata of digital images, but in Adobe Photoshop, these are translated into a format called an extensible metadata platform (XMP). Photoshop can be set up to track all adjustments made to an image file within this metadata set, and this information can be referenced through the metadata panel in Adobe Bridge®, Adobe Lightroom®, or through File > File-info in Adobe Photoshop.

One caveat of relying on metadata alone to track processing steps of an image intended for analysis is that the information is not always easy to decipher. That, of course, depends on the techniques used. Reading a levels adjustment summary in the file history is clear and easy to decipher, but if an area of interest is selected within the image with the lasso tool, *that* is going to be very difficult to interpret later. A good processing tracking system may be a hybrid of two or more tracking techniques, and metadata is a great way to corroborate other recording means. The history log must first be set up in the program to record all future adjustments made to image files.

Adobe Photoshop® History Log

The Adobe history log may be set up to record and embed everything that is done to the image in Adobe Photoshop within the file history:

1. Go to Edit > Preferences > History Log (in older versions of Photoshop, you will find the History Log at the bottom of the General tab)
2. Click to check the History Log box, turning it on, and select the following options:
   a. Save Log Items To: Metadata
   b. Edit Log Items: Detailed

Photoshop should be closed and restarted after making changes in the preferences panel.

Viewing Metadata

There are at least three ways to view metadata of the changes made to an image within Photoshop (provided it is set up to do so):

1. *Adobe Photoshop*—With the file open, go to File > File Info. To find the changes made to a file in Photoshop, look for either a *History* tab, or a *Photoshop* tab. If a printout of this data is required, it may be selected (Select all: Ctrl A for Windows, Command A for Mac), copied (Ctrl C for Windows, Command C for Mac), and pasted into a Word document (Ctrl V for Windows, Command V for Mac).
2. *Lightroom*—The library module in Lightroom has a metadata panel on the right side of the workspace.
3. *Bridge*—The default workspace for bridge has a metadata panel available on the right side also.

Aside from viewing the camera capture information and history (changes) of a file in these metadata panels, keywords and copyright information may be added to images or sets of images.

Written Notes, Word Documents, and Screen Captures

It is always acceptable to take your own notes to track changes made to a file, whether you do it with a pen and paper or with a Word document. Screen captures of steps along the way can be very helpful to illustrate what was done and to refresh your memory as required. These notes obviously don't automatically follow an image file as metadata does, so thought must be given to your policies on keeping notes with case files and disclosure procedures.

Adobe Photoshop Actions

Actions may be thought of as a "recorder" program within the larger program of Photoshop. While digital artists record and save actions to play them back on new images, recreating a desired effect, we may use it simply to track everything we do to a file as another way of keeping detailed notes. It is very similar to the old tape recorders some of us used in our youth to record music on the radio. We pushed a large red piano-key button to begin recording, another button to stop recording, another to rewind, and yet another to play it all back. Actions work very much the same way. Once a copy of the original has been opened, calibrated 1:1 for AFIS (and saved as a calibrated version), it is ready for processing. Before going ahead, it is a good idea to do some exploratory research of a file; which color channels show strong signal? Which channels harbor noise, if there is any? Can channel blending be accomplished to minimize noise before adjusting contrast (see Chapter 3)? Forming a plan or strategy beforehand makes image process recording accurate and clear to understand if it must be referred to at a later date.

If Actions has been used to record processing steps, then the palette will contain all the recording information. This may be saved as a file with your casework, both an actions file (.ATN) that can be loaded back into Photoshop to playback the steps, and as a text file (.TXT), which may be printed. The real advantage to this tracking method is that the calibrated copy of the original can be opened and the action played back, step by step, while reviewing the settings. I can't think of a better way to become re-acquainted with the processing steps of a file. It takes a little bit of organization to set up, but it quickly becomes second nature as it is incorporated into a workflow.

There are three parts to learning an Action workflow to track processing steps: *Set-up* (to record your steps), *Saving the Actions* (with case file), and *Load Actions/Playback* (to review the processing steps later) (Figure 2.2).

Setting Up Actions

1. Open the calibrated copy of the original image in Photoshop.
2. Make sure the Actions palette is open along the right side of the workspace. If it is not there, go to the Photoshop menu Window > Actions.

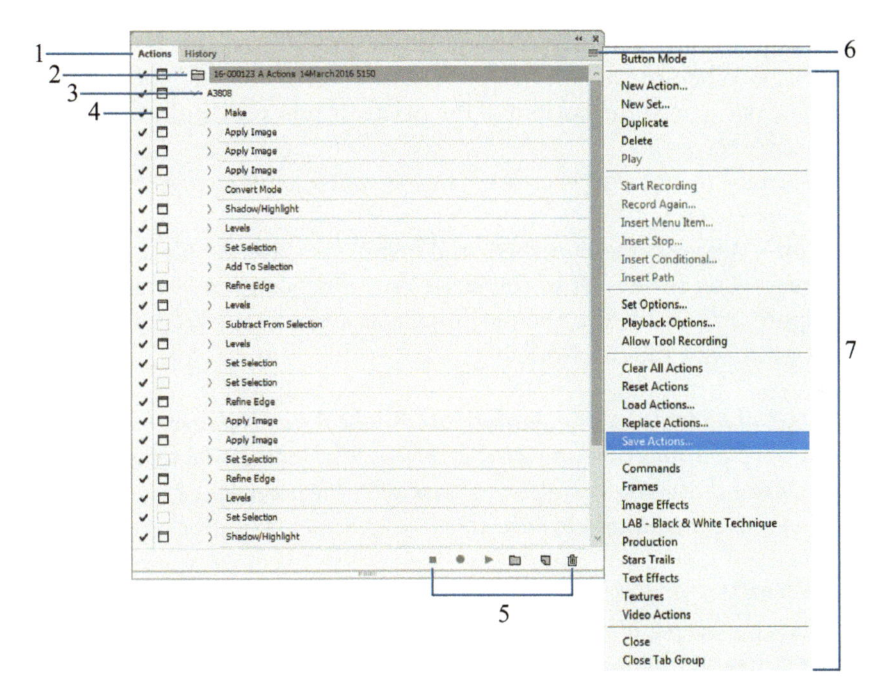

Figure 2.2 Actions palette. (1) Actions tab. (2) Set folder. (3) Action header within the set. (4) Toggle dialogue boxes on/off for playback. (5) Palette options: Stop, record, playback, new set, new action, trash. (6) Drop-down menu button. (7) The drop-down menu.

3. Clear the Actions palette by clicking the tiny drop-down menu on the top right side of the palette. You will be asked, "Delete all Actions?"—Click OK. This clears them out of the palette (but they remain on the hard drive).
4. Create a *New Set*, by either clicking the folder icon at the bottom of the Actions palette, or go to the drop-down menu and select "New Set." A small dialogue box will open, prompting you to name your set. The case number, date, and your badge number are good suggestions for naming the set, just give some thought as to what will make your workflow easiest. I also type the word *Actions* in the set name, so that when browsing the case file in windows, it is clearly identifiable as an Actions file. Once you name it and click OK, the "set" heading will appear at the top of the palette beside a tiny folder icon. This is the name of an actions folder that so far contains no actions—it is possible to save multiple actions within it.
5. If you are ready to begin processing, open the image, create a *New Action* by either clicking the paper icon at the bottom of the palette, or go to the drop-down menu and select "New Action." A small dialogue box will appear, prompting you to name your action. The specific filename of the image being processed should be entered here. This particular action is only meant to go with the corresponding image. When the action has been named, click *Record*. Notice that the circle icon at the bottom of the palette is now red, indicating that you are recording.
6. Process—*Notice each processing step is added to the sublist below the Action Heading.*
7. Stop—When you are finished, and before you save the file, click the square icon to the left of the circle *recording* icon (see Section 5 of Figure 2.2). Then save the new image as a processed version of the original with the casework.

Note: Not all tools are compatible with Adobe Actions; dodge, burn, painting in a selection mask or any kind of brush work cannot be recorded. It is recommended that Actions are played back after the processing is completed to ensure total documentation of each step.

Save Actions

8. Click the *Set heading* to select/activate it. If any other step within the palette is highlighted (active), you will not be able to save; grayed-out save options in the menu may indicate that the "Set" is not selected.
9. Go to the drop-down menu for Actions (top right of the palette), and choose save Actions. A navigation window will open; browse to save in case folder. Notice that the file format is .ATN.

10. To save as a text file: Hold the Ctrl and Alt keys down (Command and Option keys on Mac), and go to the drop-down menu again. Choose save Actions. Again, navigate and save to the desired folder, only this time the file format being saved is .TXT. *Note: Extra notes may be written into the appropriate place in this text document. For example, if the image has been taken into another software program as part of the processing step, it may be documented here.*
11. Additional security—All files may be made Read-Only (calibrated, processed, actions). In Windows, right-click > Properties > Attributes (click Read-only). Secure an extra copy of the folder in an alternate location (redundant backup).
12. From the drop-down menu, *Clear All Actions* may be selected—the Actions files have been saved but may be cleared from the Actions Palette.

Playback Actions

You may wish to ensure that your Actions play back correctly on your calibrated copy of the original, or you may need to refresh your memory on processing steps later.

13. Re-open the calibrated image you started with (close any other images open in Photoshop).
14. Go to the Actions drop-down menu (top right of palette), and click *Load Actions.*
15. Navigate to the folder where the Actions are saved with the case files. Click and load.
16. Click on the Action header that corresponds with the filename you are reviewing.
17. Click the square box to the left of the Action header to toggle *dialogue boxes on* (see Section 4 in Figure 2.2); this will make the program stop at each step that includes a dialogue box so that you can review the settings. It will not move on to the next step until you click OK.
18. Go to the drop-down menu (top right of palette), and click playback options. Performance defaults to "accelerated." Change to "Pause for 1 second."
19. Click the triangle Play button (bottom of palette). Each step taken will be played back one at a time. At each dialogue box, the program will stop and wait for you to click OK before continuing to the next step.

Adobe Photoshop Adjustment Layers

Adjustment layers are another great way to track processing steps. As with Actions, not all tools are compatible with this tracking method. Apply image and calculations cannot be tracked with adjustment layers, but channel mixer and black and white filter can (covered in Chapter 3). Review the tools that you know and understand; use the tracking method that works best for those tools.

Multiple adjustment layers may be created for making color and tonal adjustments to an image (curves, levels, channel mixer, black and white filter, etc.). They will affect the base image below them without changing any of its pixel values. The effects of these layers can be turned on or off, and the settings for each adjustment can be reviewed. A layer mask is automatically created for each adjustment layer, allowing for selective editing of the effect.

Converting to a smart object allows for further nondestructive editing by allowing filters to be used such as Adobe Camera Raw filter, and unsharp mask. These filters may be applied to any RGB or grayscale image, even JPEG, and will be listed with settings in a layer below the base image (smart object). The layered image may be saved as either a TIFF or PSD file, but be aware that multiple layers increase the file size significantly, and if this file must be shared with others, they may not be able to view the file with its layers. A flattened version may be presented as the processed image, as long as the original is preserved, and the adjustment layers are preserved for an expert of comparable training to review and duplicate as required. These procedures should be clearly outlined in a department's standard operating procedures (SOP).

Smart Objects

A smart object can be a raster image (photo) or a vector layer on which nondestructive editing may be applied; all the original pixel content is protected and preserved. Filters such as Adobe Camera Raw filter and unsharp mask can be applied without making any changes to the original pixel data. These filters will be listed similarly to an adjustment layer, only *below* the image layer.

Adobe Camera Raw®

While it is possible to open a JPEG in Adobe Camera Raw (ACR) if it is in RGB or grayscale mode, be aware that an 8-bit image cannot truly be converted to a 16-bit image here. Other adjustment features in Adobe Camera Raw filter can be applied to a JPEG file, however, and saved as a layer with settings intact, provided the image is first converted to a smart object. With the image open and active, go to Filter > Convert for Smart Filters. A warning will pop up reading: "To enable re-editable smart filters, the selected layer will be converted into a smart object." Click OK.

The layer is now called "layer 0" instead of "background," and the small icon in the bottom right corner of the layer thumbnail indicates that it is a smart object. ACR adjustment settings will now be saved as an "adjustment layer" under the smart object. Other filters such as smart sharpen and unsharp mask may also be applied and will appear in the filter list below the image along with the settings used (Figures 2.2 and 2.3).

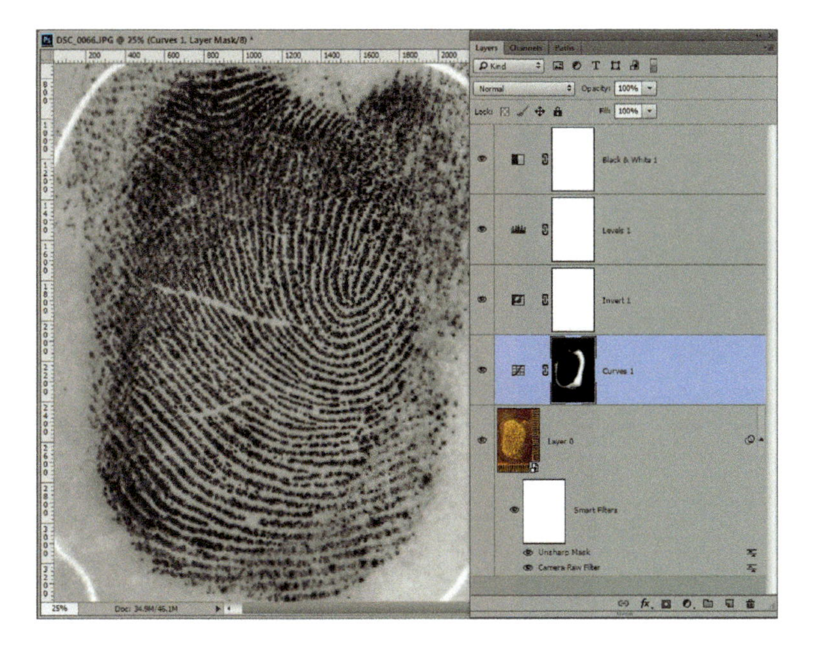

Figure 2.3 Layer palette of a smart object with Adobe Camera Raw and sharpen filters applied (listed in the layer below the smart object) as well as several adjustment layers.

Creating an Adjustment Layer

Three ways to create an adjustment layer in Photoshop:

1. Click the Adjustments tab to activate the palette containing adjustment tool options. This palette lives above the Layers palette by default, but if it is not there, go to Menu, Window > Adjustments. There is a small icon representing each adjustment tool—hover your mouse over each one so its name is displayed above them. As soon as an icon is clicked, a properties box for that tool opens, and immediate adjustments may be made. An adjustment layer for this tool is automatically created above the base image you are working on.
2. Click the Adjustment Layers icon at the bottom of the Layers palette (a circle half-filled with white and half-filled with black), and you will be presented with a list of adjustment tools that you may choose from to apply to the base image. A new adjustment layer will be instantly created over the base image including a layer mask (indicated by a white rectangle mask beside the thumbnail in the layer bar).
3. Go to the menu, Layer > New Adjustment Layer, and choose the desired adjustment tool from the drop-down list. A New Layer dialogue box will open for that adjustment tool. Click OK, and the Properties Box will open for that tool.

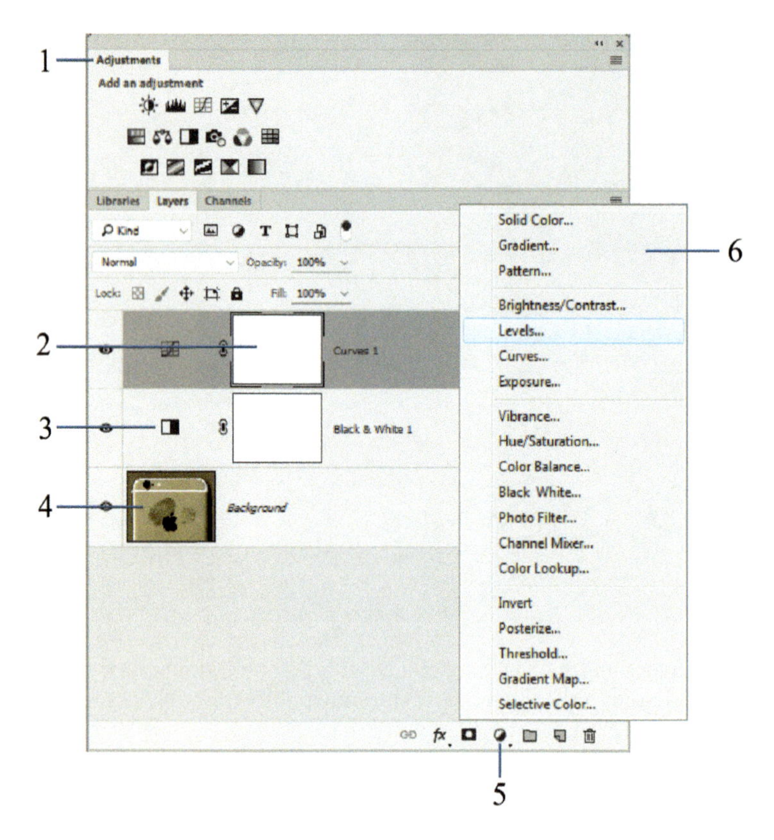

Figure 2.4 Layers and Adjustments palettes. (1) Adjustment palette tab. (2) Layer mask thumbnail. (3) Adjustment layer icon. (4) Base layer image icon. (5) Create new adjustment layer icon. (6) Adjustment tools menu.

Make adjustments in the properties box; the settings will be stored in the layer above the base layer for future reference (Figure 2.4).

Multiple adjustment layers may be added to process an image, and the entire composite image can be saved—settings are stored in each one. To edit an adjustment layer, double-click the layer thumbnail; the properties box will open to the settings used. Further changes may be made to these adjustments at that point, so if it is desirable to preserve the original settings, save the composite file as read-only (navigate to Case Folder in Windows, select the file, right-click, go to Properties, click Read-Only).

Adjustment layers may be toggled on and off to view the effects before and after. Click the eyeball icon to the left of the layer thumbnail—eyeball on, no eyeball, off.

About Adjustment Layer Masks (See Figure 2.4, Feature 2)

Each adjustment layer is created with a white layer mask thumbnail to the right of the tool icon, showing all the adjustment effect to the entire image.

It is possible to mask out some of the effect to portions of the image. This requires painting the mask with black to block the effect 100% in those areas (painting with a shade of gray masks out the effect by varying degrees—the lighter the shade of gray, the more the effect shows. The darker the shade of gray, the less the effect shows).

Editing a Layer Mask

Click inside the white *Layer Mask Thumbnail* to activate it (the mask box will be highlighted to show it is active). Use a black brush to paint "out" the adjustment tool effects. Paint with white to paint effects back in.

Note: This cannot be tracked with Actions as a process recording method (Figure 2.5).

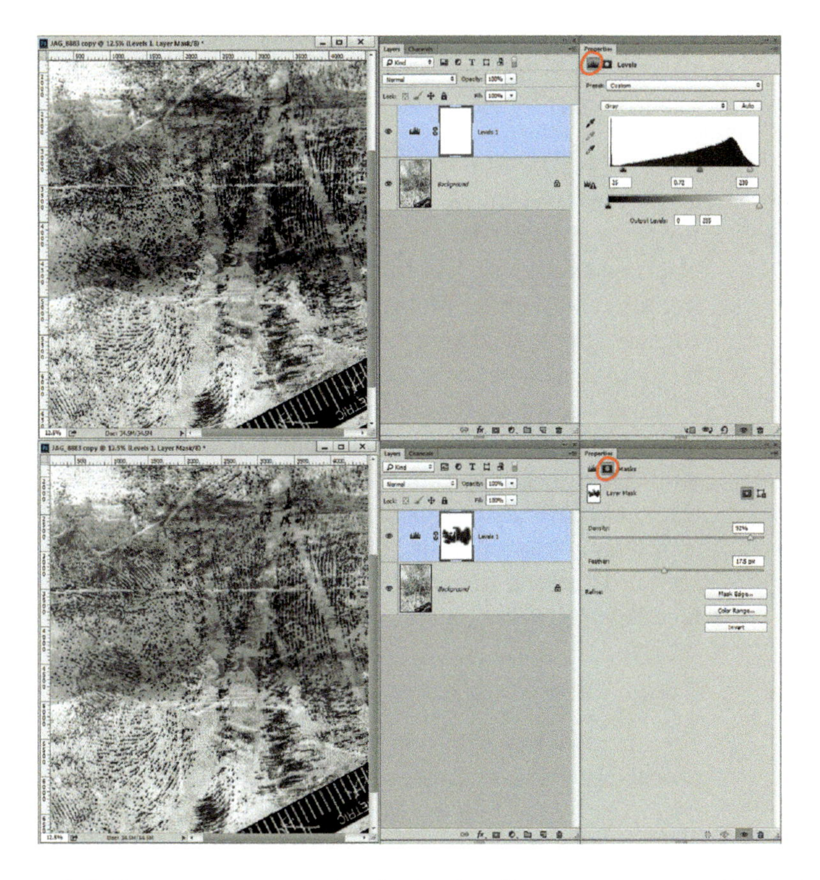

Figure 2.5 Top row: An example of a levels adjustment layer applied to base image. Note the white mask in the layers palette shows all effects. The properties palette (top right), shows the levels histogram settings. Bottom row: The same image and levels settings, only the mask has been edited with a black brush (center). The properties palette now offers mask editing options. (Image courtesy of Jason Goodfellow, York Regional Police.)

Using Brushes to Edit Masks

Select your brush from the toolbox, and make sure that your foreground color is black. Begin editing the mask. Change brush color to white if too much black has been painted in.

Tip: *Use a feathered brush to blend well. When the brush tool is active, go to the tool options palette across the top, and click the "Click to Open the Brush Preset Picker" to set the hardness to 0* (Figure 2.6).

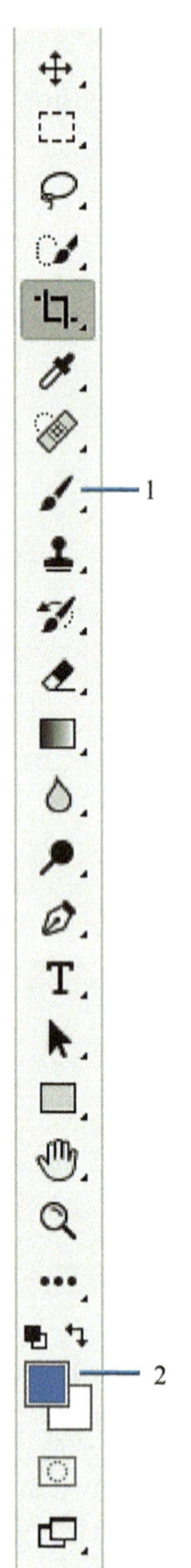

Figure 2.6 Toolbox. (1) Brush tool. (2) Foreground and background color swatches.

Keyboard shortcuts for brushes:

- "d"—Sets the foreground and background colors to their default settings (black for foreground and white for background)
- "x"—Switches the foreground and background colors
- Left and right square brackets—Make the brush size smaller or larger as you work

Property Palette—Mask Editing Options

Click the layer mask to show mask editing options in the Property palette. The density slider will lighten or darken the black masking that has been created to show the adjustment in those areas, either more or less. The feather slider allows the edges of the masked area painted in to be softened by a width of pixels indicated on the slider.

Adjustment layers continue to be stacked on top of each other, and if the image has been converted to a smart object to allow for filters to be applied, then the filters are stacked below the image layer. Save this image as a PSD or layered TIFF, and lock it as a read-only file so that the original settings are preserved. Flatten it (menu > Layer > Flatten) and save as a flattened copy. It will be a very large file.

Adobe Photoshop Notes

The note tool in Photoshop may be used to keep notes of image processing steps, and these notes can be attached to the image itself. Go to the tool box, and click and hold the eyedropper until a menu pops out with more tool choices. One of these is the note tool. Click on it, and your mouse pointer turns into a note icon. In the Option bar for this tool (across the top of the desktop under the menu), there is a place to insert the author of the note. Name, badge number, and date can be documented here. A note color may also be selected. Click anywhere within the image to drop a note; a colored note icon will be placed within the image (does not print when image is printed), and a note panel opens with cursor ready for text input.

To delete a note, select the notes icon in the tool box, then click the trash can icon at the bottom of the notes panel, or click the *clear all* button in the options panel to delete all notes attached to the image.

Notes will not be printed with the file when it is printed, nor will the icon show on the printed image. To print notes separately, copy and paste into a Word document.

Third-Party Software Image Process Recording

Third-party software solutions can be useful tool for storing, securing, locating, and controlling digital image evidence gathered at crime scenes. These software solutions allow forensic photographers to upload their crime-scene and lab images through an interface that is easy to use, and they manage the storing of originals in a separate location (redundant backup), further allowing browser-based viewing of cases and digital evidence, along with comparisons of originals and processed copies. It can authenticate original image captures, allowing processing of only copies. Software may store and track information (such as camera EXIF data), history of processing, and an audit trail of all access to the digital asset.

There are many digital asset management systems available to choose from, but it requires a little research to find the right fit for your needs.

RAW File Formats and Image Processing

RAW images are proprietary file formats created by the camera manufacturer, offering the photographer the most control over the processing of their images. A common complaint about RAW images is that they look flat and dull as compared to their JPEG counterparts, but it must be understood that adjustments need to be made in software such as Adobe Camera Raw to produce a good image.

Shooting RAW offers great advantages to the forensic photographer/processor, particularly in the context of images intended for analysis.

1. RAW is considered to be a digital negative. They are uncompressed (lossless), original data is retained without *any* processing applied, and they cannot be overwritten. The image results of any adjustments made in RAW editing software such as Adobe Camera Raw must be saved as another image format such as JPEG or TIFF, while the RAW file itself remains original. When making adjustments to the file in ACR, a sidecar file containing the adjustment settings is created

automatically, saved with the same name and in the same location as the original RAW with an XMP extension. Both the RAW file and the XMP file should remain together, as these are the first processing steps toward clarifying an image intended for analysis.

2. RAW images can have greater dynamic range (14- to 16-bit depth), allowing for the possibility of gleaning more image detail in the shadows and highlights. This results in a better starting place from which to begin further processes, with as much detail as possible maintained in the highlights and shadows.

   Conversely, when shooting in JPEG, the loss of original data and quality can be attributed to three criteria:

   a. Bit depth is compressed to 8 bits. This compresses the color and brightness values in an image, potentially losing crucial detail. This is most noticeable in the darkest shadows and brightest highlights of an image. Images photographed with a forensic light source, for example, often consist of areas that are overexposed and "blooming," while other areas of the print fluoresce subtly and reside in the dark shadows of the image. The very high dynamic range of these images is sometimes far too drastic, and if captured in JPEG format, detail may be permanently lost.
   b. JPEG is automatically compressed, and original pixel data is thrown away permanently. Compression algorithms are very sophisticated these days and do a great job creating smaller JPEG files that can be easily shared, emailed, or posted on the Web; however, this is not recommended for images intended for analysis. Compression not only results in loss of original pixel data, but some artifacts and noise are introduced into the image in the process, and those disturbances are most noticeable around the edge details. And what is a fingerprint image, if not a series of edge detail?
   c. The analog to digital converter or processor (A/D converter) applies numerous processes to a JPEG image before saving it to the memory medium. It automatically adjusts contrast, color balance, white balance, and even sharpening, which may result in clipped tones of the highlights and shadows. These things all contribute to loss of original data at capture that cannot be undone.

The Hitch

As mentioned, RAW files are proprietary formats created by the camera manufacturers, and each one creates their own with specifications that are not publicly available. Consideration for long-term archival must be considered, and it may offer complications when sharing across complex workflows.

The creation of the digital negative (DNG) format is an answer to that problem. It is a publicly available format for proprietary format RAW files.

Many software manufacturers currently support DNG, and some camera manufacturers are beginning to develop direct support for DNG format.

Adobe offers a free DNG converter that allows users to convert the RAW files from over 350 cameras to a DNG format. This may be useful when using older versions of Adobe Photoshop and shooting images with newer cameras that are not supported in the older software. Find the free converter on Adobe's download page: http://www.adobe.com/support/downloads/.

Introduction to the Adobe Camera Raw Dialogue Box

Opening RAW images for image processing requires making some adjustments in the Adobe Camera Raw dialogue box as the first step in most processing plans. This is an excellent opportunity to lighten shadows and subdue highlights to visually maximize any detail that may be residing there. Counterintuitively, preadjustments made in ACR often (but not always) result in lower contrast images that are a much better starting point to make existing information easier to visualize. Adjusting contrast later, near the end of the process, after minimizing noise distraction is often a more effective strategy.

The ACR tools as numbered in Figure 2.7 are briefly explained below; keyboard shortcuts are given in brackets beside each too.

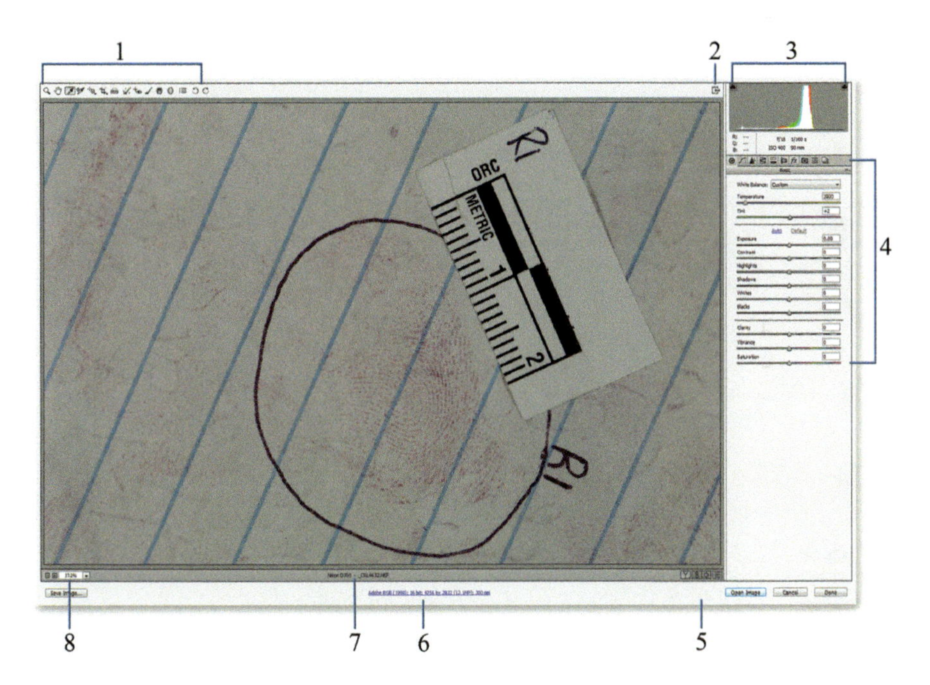

Figure 2.7 Adobe Camera Raw dialogue box. (1) Camera Raw tools. (2) Toggle full-screen mode. (3) Histogram. (4) Adjustment panel controls. (5) Cycle between before and after views. (6) Workflow options. (7) Filename. (8) Select zoom level. (Image courtesy of York Regional Police. All rights reserved.)

1. *ACR Tools* (Figures 2.8 through 2.22)

Figure 2.8 Zoom (Z).

Figure 2.9 Hand (H): Use the Hand tool to scroll within an enlarged preview of the image.

Figure 2.10 White balance tool (I): Use the White Balance eyedropper to click on an area of the image that is supposed to be neutral, to eliminate color casts.

Figure 2.11 Color sampler tool (S): Place as many as nine color sampler points in your image to "live-track" the color values for those points as you work.

Figure 2.12 Targeted adjustment tool (T): Select very specific areas within the image and adjust the curves (parametric curve), hue, saturation, luminance, and grayscale mix to create black and white images.

Figure 2.13 Crop tool (C): Crop your image as you would in Photoshop.

Figure 2.14 Straighten tool (A): Select the Straighten tool and drag along the line that is intended to be horizontal. Double-click to straighten automatically.

Figure 2.15 Spot removal (B).

Figure 2.16 Red eye removal (E).

Figure 2.17 Adjustment brush (K).

Figure 2.18 Graduated filter (G).

Figure 2.19 Radial filter (J).

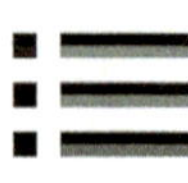

Figure 2.20 Open preferences dialogue (Ctrl K).

Figure 2.21 Rotate counterclockwise (L).

Figure 2.22 Rotate clockwise (R).

2. *Toggle Full-Screen Mode* (F)
   The ACR dialogue box may be operated from a smaller window within the Photoshop desktop, or it may fill the monitor screen.
3. *Histogram*
   The histogram plots the tonal values of the image from blacks (far left), to midtones (center), and finally to highlights and whites (far right). Learning to read a histogram can help both with shooting images proficiently and editing them (more on histograms in Chapter 6). In the top corners of the histogram box, there are arrow buttons. The button on the top left will toggle on/off the shadows that are being clipped to pure black. The button on the top right will toggle on/off the highlights that are being clipped to pure white. Adjustments can be made to rectify any severe clipping that may result in lost detail. Note that some images may have naturally occurring clipped highlights; specular highlights such as the sunshine glinting off water or reflecting from glass may be acceptable, for example.
4. *Adjustment Panel Controls*
   The panel tabs across the top of all these sliders open up new adjustment possibilities. The first tab is Basic camera adjustments, and for the purposes of this book, our RAW adjustments will be made here. Moving left to right, the panels are: Basic, Tone Curve, Detail (sharpening and noise reduction), HSL and Grayscale, Split Toning, Lens Corrections, Effects, Camera Calibration, Presets, and Snapshots.
5. *Cycle between Before and After Views* (Q)
6. *Workflow Options*
   Lists the color space, bit depth, pixel dimensions (file size), and pixels per inch, but this is also a link to a Workflow Options panel where this information may be set for opening every RAW file (Figure 2.23).

   Options may be customized to tell the software how to open all images in ACR. The workflow options box offers some customizable features:

   - Color space—Select the desired color space for opening all images, and choose bit depth. It is recommended to open all images for analysis as 16-bit depth files to allow for the best chance to make existing but faint detail more visible. The last step of processing may be to convert the file to 8 bits so that it is compatible with printers and other software systems such as the Automated Fingerprint Identification System (AFIS).
   - Image sizing—Specific image dimensions and file size may be set for opening files but not recommended for forensic purposes.

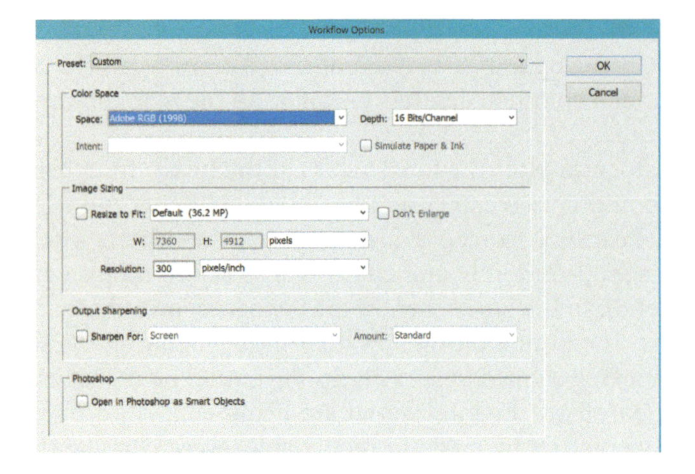

Figure 2.23 Workflow options.

- Output sharpening—Automatic sharpening for output files may be set here, but again, it is not recommended for images intended for analysis; those adjustments are arguably best made much later in the process.
- Photoshop—Can open all images automatically as smart objects, if that is part of your workflow.

7. *Filename of Image Open in ACR*
8. *Select Zoom Level*

More on ACR's Adjustment Panel Controls

Basic camera control adjustments may be adjusted in the order that they are listed in the panel. White balance refers to color temperature, describing the warmth or coolness of the image lighting. This setting defaults to "As Shot", but you may click the drop-down arrow to specifically describe the lighting conditions for an image: auto, daylight, cloudy, shade, tungsten, fluorescent, flash, and custom. As soon as the temperature and/or tint sliders are manually adjusted, the color balance changes to "custom." Image color balance may be achieved if there is something in the image that is known to be a neutral black, gray, or white by selecting the white balance tool (gray eyedropper) and clicking in the neutral area.

The image tonal adjustment controls start with the exposure slider. Image exposure may be increased by five stops or decreased by five stops. While an incredible amount of detail may be recovered from the darkest shadows or highlights, this depends on their existence in the first place. A very badly underexposed image may not have captured all detail within the shadows, and even with a 16-bit image, it may be unrecoverable because it is not even there. The point is that although the exposure slider in ACR may be able to improve the exposure and detail of an image, it is still important to try to capture the image correctly from the outset to ensure the best quality output possible.

Tip: *It cannot be stressed enough to achieve the ideal exposure at the moment of capture for the image intended for analysis. For example, a photographer may perfectly expose the scale on a black or dark object, but the fingerprint itself is grossly underexposed. While the exposure can be increased in ACR, and the shadows may be lightened to reveal the detail of the fingerprint, there is invariably a great mess of noise introduced into the corrected version because noise lives in the shadows. Also, if the image has been too underexposed, the detail in the darkest shadows may not have even been captured. As soon as the contrast is increased at a later point in the process, that noise starts to make the whole image look like oatmeal. Try exposing for the fingerprint, even if the scale is a little overexposed (provided that the detail in the scale is still discernable for calibrating).*

The contrast slider is self-explanatory. Consider going easy on the contrast at this point—the first goal is to ensure that all the image detail is visible. Increasing contrast at the outset can put details in highlights and shadows at risk; contrast is often best left to be among the last processing steps in a processing workflow.

Highlights and shadows sliders can be extremely beneficial to some images. Fluorescent photography, for example, presents some unique challenges when clarifying fingerprint detail; some ridges may be overexposed and bloom brightly, while other areas fluoresce faintly and are almost lost within shadow. Highlights and shadows sliders may even out these tones to help visualize those details more easily. Blooming bright ridges or blotches may be toned down by darkening just the highlights in the image. Dark shadows threatening to engulf those faintly fluorescing ridges may be lightened making those ridges easier to visualize.

The whites and blacks sliders adjust clipping in the whites and blacks. Drag the whites slider to the left to reduce clipping in the highlights. Drag the blacks slider to the left to reduce clipping in the shadows.

The clarity slider increases contrast, with the greatest effect to the midtones, reducing the risk of losing detail in the highlights and shadows.

The *vibrance slider* selectively increases the saturation of low-saturated colors but with a reduced effect on more highly saturated colors. It is found to be a useful tool in portrait image processing, preventing oversaturation of skin tones while boosting other colors.

The saturation slider adjusts the saturation of all image colors equally.

RAW Highlights and Shadows

Figure 2.24 is a RAW image of a fluorescing fingerprint, opening automatically in 16-bit mode, as workflow options have been set to do so. While intuitively one might be tempted to increase contrast between the ridges and the furrows in a fingerprint, this example illustrates when high contrast is just too much, at least at this point. Instead, by lightening shadows and darkening highlights, the resulting lower contrast image offers a better starting place for enhancing ridge detail. Increasing contrast of a fingerprint image is one of the last steps in image processing and requires a gentle hand at that.

Figure 2.25 illustrates a better starting point. Exposure has been increased almost a full stop, contrast has been reduced, shadows lightened, highlights darkened, blacks lightened, and clarity has been increased. The result is a lower contrast image (effects similar to high dynamic range HDR merge), revealing more of the existing fingerprint ridge detail that was previously difficult to visualize, particularly in the shadows. Figure 2.26 illustrates a zoomed in version of both images for comparison; ACR default settings are featured on the left, while the low contrast settings are displayed on the right. This image will be revisited for further processing with other contrast tools later in this book.

As previously mentioned, once a RAW file has been opened in ACR, a sidecar file is created (.xmp). This file should be stored with the original RAW file in an image processing folder, and it is suggested that it be made a *read-only* file to protect the settings from being inadvertently altered at any later date (right-click, choose Properties, and click the "Read-only" box).

Figure 2.24 ACR dialogue box showing image with default settings. (Image courtesy of York Regional Police.)

Figure 2.25 ACR dialogue box showing image with low contrast settings. (Image courtesy of York Regional Police. All rights reserved.)

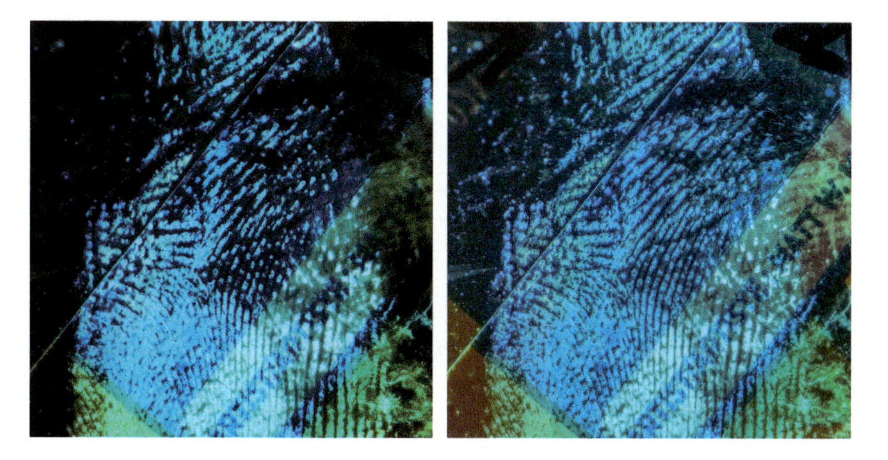

Figure 2.26 Image comparison between Figures 2.24 and 2.25. (Image courtesy of York Regional Police. All rights reserved.)

Image Calibration

Images captured with a camera on a tripod or even on a copy stand are not necessarily 1:1 (life-size when printed), so images must be scaled using software before they are ready for AFIS (Automated Fingerprint Identification System) submission.

Calibrating an image 1:1 in Adobe Photoshop is very quick and easy. It is the second step (after ACR adjustments) to most forensic processing workflows, and while it is still recommended to be done on a working copy of the original, it does not change any of the pixel data in the image whatsoever. Calibrating an image 1:1 changes the instructions for printing the image's pixels on a page, telling the printing software how much paper area the print is going to cover, yet the pixel dimensions of the image remain exactly the same as the original. For example, our Nikon cameras (at the current settings) are producing images that are 6,000 by 4,000 pixels. The default physical dimensions, if printed, are 20 by 13.333 inches at 300 pixels per inch. A fingerprint filling the frame will be far larger than real life if printed at this size. Calibrating this image 1:1 simply stores the new physical dimensions without resampling (no added or subtracted pixels) so that the printer prints at the new physical size as instructed (2.977 by 1.985 inches at 2,015.211 pixels per inch) while maintaining the original dimensions of 6,000 by 4,000 pixels.

Just to go ahead and state the obvious, a little forethought is required at the image capture stage. Your subject, whether it is a fingerprint, footwear impression, bite mark, bruising, tire tread, or tool mark, must be photographed as follows:

1. A forensic-quality scale must be included in the image and should not cover any part of the subject (fingerprint, bruise, etc.).
2. Both the subject and the scale should be 90 degrees to the camera sensor plane (or film plane). Any angled or curved scales will reduce the accuracy of your calibration attempts later.
3. The scale must be placed alongside and on the same plane as the subject; do not allow your scale to be placed behind or in front of your subject, as this will cause inaccuracy in sizing.

Resolution Is a Three-Headed Monster

The image size dialogue box describes the dimensions of any image. Information such as the pixel dimensions, physical dimensions such as width and height, as well as how many pixels will fit into an inch are all there. The trick to image calibrating 1:1 is to work with the dimensions of an image without adding or subtracting any pixel data (called *resampling*); the width, height, and pixels per inch (ppi) must be linked together. This means if the width value is manually changed, the height and the resolution will change too. If the height is manually changed, the width and the resolution change. Therefore, if the resolution is changed manually, so will the width and the height. Provided the *Resample* option is turned off, the relationship between these factors will remain the same; they are inextricably linked, and there are an infinite number of sizing combinations possible. Only *one* of them will result in an image calibrated 1:1. The cool thing about this understanding is that if we can find any one of those three criteria that will result in a true, life-size image, then the other two criteria values will fall into line, as discussed below.

Steps to Calibrate Image 1:1

1. After making adjustments in ACR, open image.
2. Set-up: If rulers are not already lining the left and top side of the image (see Figure 2.27, 5), enable rulers in View > Rulers; simply click to toggle them on (with checkmark). They will stay on for all future images opened until unchecked.
3. Set-up: Select the Ruler tool from the Photoshop toolbar (*Note: The ruler tool may be found under the eyedropper tool on the toolbar. Click and hold the eyedropper tool to view a hidden menu of extra tools;*

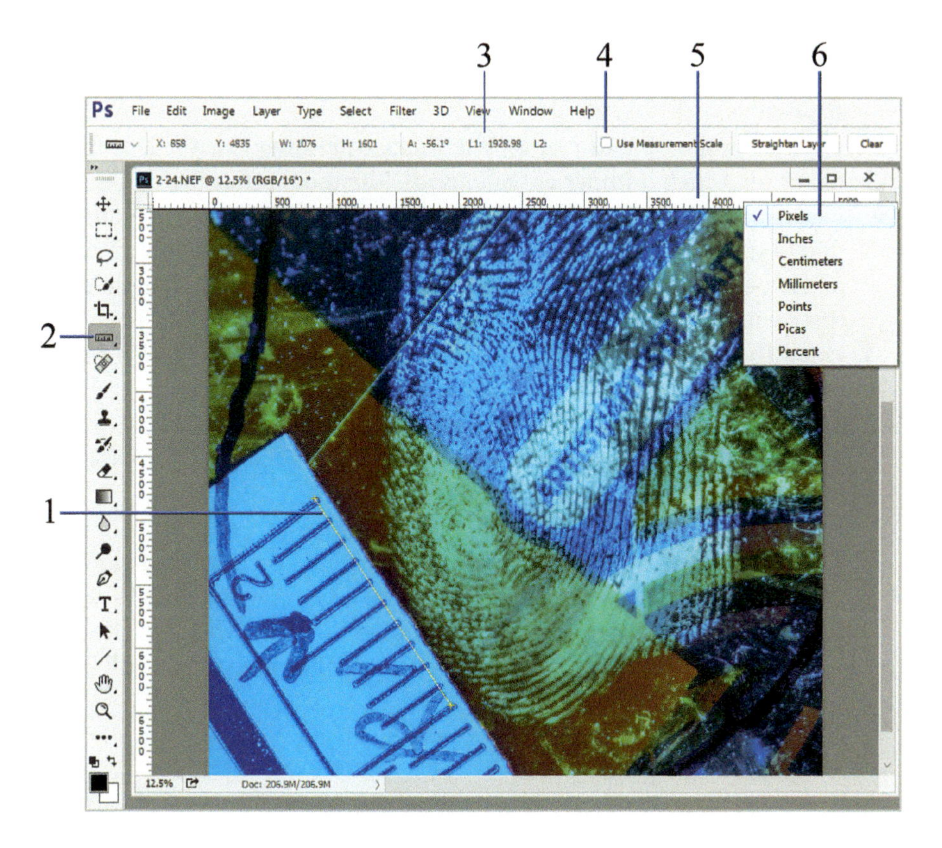

Figure 2.27 The ruler tool and options—Preparing to Calibrate. (1) Measurement drawn across 1 centimeter of the scale. (2) Ruler tool. (3) Line 1 value. (4) Use measurement scale option. (5) Rulers showing. (6) Unit options available (right-click within the ruler area). (Image courtesy of York Regional Police.)

one of these is the ruler tool). Ruler tool options bar will be displayed across the top of the workspace just under the menu bar.

   a. Ensure the "Use Measurement Scale" option is *unchecked* (see Figure 2.27, 4). It is not necessary to use the measurement units in the program settings.

4. Set up: Place cursor inside the ruler area surrounding the left and top side of the image, and right-click. From the drop-down menu, choose Pixels (see Figure 2.27, 6). This sets the ruler tool to measure how many pixels there are in any line drawn. Once it is set to pixels, it will stay that way until manually changed.
5. Using the ruler tool, measure 1 centimeter across the scale in the image or 10 millimeters (see Figure 2.27, 1). Click at the start of the centimeter, draw to the 1-centimeter mark, and release. It is advisable to zoom in to be more accurate with measurement. The line may be drawn from the center of the ruler markings or edge-to- edge (as long as you use the same edge on both ends for accuracy). If the image contains an imperial scale, then measure 1 *inch* as accurately as possible.
6. There is now a line drawn on the image spanning exactly 1 centimeter (or 1 inch if imperial scale used). The tool options bar, L1 (see Figure 2.27, 3) displays the number of pixels across that line. Because that line spans 1 centimeter in the image, it is now known how many pixels per centimeter it takes to make this image 1:1. Make note of this value, or write it down.
7. Open the image size dialogue box (Image > Image Size). Turn the resample option *off.*
   a. Image size—Reports the file size of the opened image.
   b. Pixel dimensions—Reports how many pixels there are for width and height.
   c. Fit to: Leave it set at the default—Original size.
   d. Width, height, and resolution—These values are displayed along with the option to change the units of measurement if desired. To the left of these, a bracket and a link will indicate which of these are linked together. When resample is off, all three values will be linked together.
   e. Resample—Make sure it is *unchecked*. Again, to clarify, *resample must be off.*
8. Beside Resolution, click the drop-down menu for pixels per inch, and change it to pixels per centimeter. Now enter the L1 value (the value representing how many pixels are in a centimeter) from our previous measurement. Of course, if the scale is imperial, then leave this at "pixels per inch," and enter the L1 value (Figure 2.28).
9. Click OK. Done.

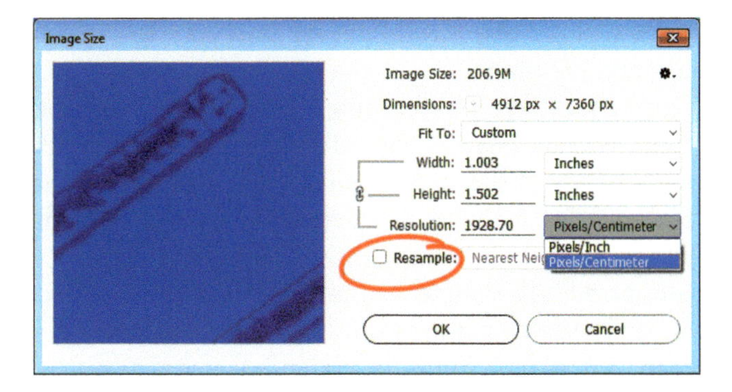

Figure 2.28 The image size dialogue box. Make sure "Resample" box is not checked. Pixels per inch must be changed to pixels per centimeter when measuring with a metric scale.

To check your work and confirm that the 1:1 calibration was successful: Right-click in the ruler area, change the measurement units to centimeters, select the ruler tool, and draw a 1-centimeter line on the scale. L1 should read 1.00 if the image has been correctly scaled.

Note: There is sometimes a glitch in the older software that prevents L1 from reporting this value correctly, so if it is not reporting as expected, click and immediately unclick the "Use Measurement Scale" option (see Figure 2.27, 4). If it is the glitch, then this should fix it. If not, you may need to try again. Print image and measure with a scale to confirm accuracy.

About Read-Only

If a third-party image management system isn't being used, it is a good idea to make the calibrated image a read-only file. This prevents you or anyone else from inadvertently overwriting the file and gives it a little added protection.

In the folder, right-click Calibrated File, go to Properties, and check the Read-only option. This file cannot be overwritten without changing this setting in the properties box.

Calibrating an Image 1:1 without a Scale

Obviously, we do not wish to advocate such an omission, but on occasion, we do receive fingerprint processing requests where the scale is absent in the photo, and the property is not accessible for a re-shoot. Sometimes there is no reference with which to calibrate an image, and that is the end of that, but occasionally, there are instances where other "known" distances can be measured, and the image can be calibrated based on those measurements.

Figure 2.29 is a photo of a fingerprint on an iPhone 6. The photographer forgot the scale, and the phone is no longer in our possession. Because of the overall photos taken and the photographer's notes, we know what model of iPhone it is, and there are identifying features that can be measured that are common to all iPhones of the same model. Once a replica of that model is acquired, it is decided that the measurements taken will be from the Apple logo itself—bite mark tip to bite mark tip.

Note: It is important to remember that the known distance you are measuring must be on the same plane as the subject you wish to replicate one to one. Anything in front of or behind the reference you are measuring will NOT be one to one.

1. Measure a known distance on your subject very carefully. This is a physical measurement of our substitute iPhone. 6 Figure 2.29 illustrates the area measured across the bite mark of the Apple logo (6.5 mm).
2. Open image.
3. Select the crop tool from the toolbox.
4. Crop the actual known distance in the image (in this case, it is from one side of the apple's bite mark to the other). It may be necessary to rotate the crop box, allowing the height (or width) of the new image to be the known distance being measured. Zoom in (Ctrl +) to make this crop as accurate as possible. Hit Enter to commit to the crop. You now have a small cropped version of your image. Do not be concerned that you have cut out some essential image information; the image will not be left this way (Figure 2.30).

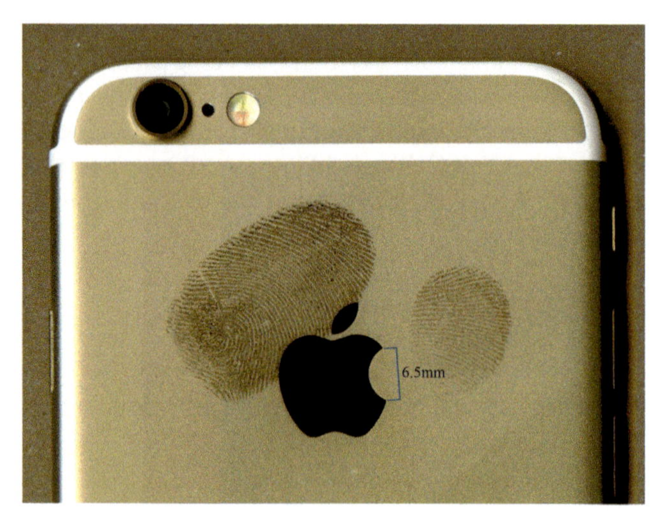

Figure 2.29 Fingerprint on iPhone 6—No scale. Measurement taken from a replica iPhone, of bite mark in Apple logo (6.5 mm).

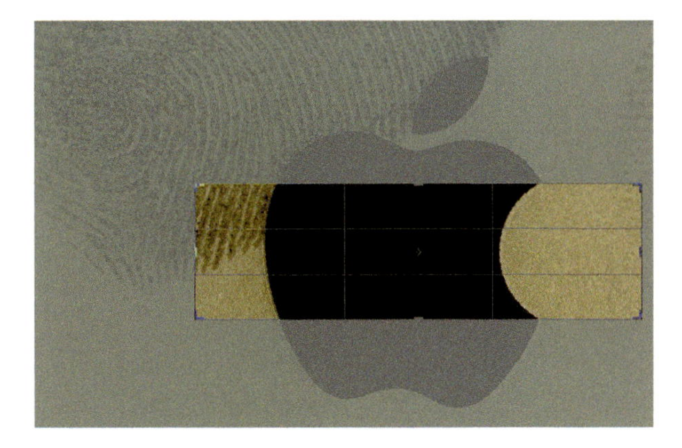

Figure 2.30 Cropping the known distance on the image (rotate the crop box by holding the cursor outside the box until it turns into curved arrows. Then click and drag box into position).

5. Go to image size dialogue box (Image > Image Size):
   a. Make sure that resample is *off*.
   b. We know that the height of this cropped image is now 6.5 mm. Enter height as 6.5 mm, and the width and resolution (pixels per inch) will change automatically.
   c. Select and *copy* the new value in the resolution box (ppi).
   d. Click *cancel* (the resolution known to make this image 1:1 has been copied for next step).
6. Go to Edit > Undo (Ctrl > Z for PC, Command > Z for Mac). This undoes the crop and takes us back to the full image.
7. Go to image size dialogue box once again (Image > Image Size):
   a. Make sure that resample is *off*.
   b. Enter the new resolution value for pixels per inch.
   c. Click OK.

This image is now calibrated 1:1 without using a scale. Remember that this is dependent on finding a known measurement in the image that is both on the same plane as the fingerprint (or footwear, bite mark, tool mark, etc.), and the image must be 90 degrees to the camera back.

Review Questions

1. What three groups may be referenced to assist with policy, best practices, standards, and guidelines for ensuring the integrity and acceptance for forensic digital images?
2. Name five ways to assist with maintaining the integrity of forensic images.

3. What file format/s are best used for capturing images intended for analysis?
4. Name five different methods of tracking the processing steps of an image intended for analysis.
5. What file format is best suited for maintaining image integrity and allows the most flexibility in clarifying image detail?

References

1. Overview of SWGIT and the Use of Imaging Technology in the Criminal Justice System (Version 3.3 2010.06.11), available at https://www.swgit.org/documents/Current%20Documents
2. SWGDE Scientific Working Group on Digital Evidence, available at https://www.swgde.org/
3. NIST and the Organization of Scientific Area Committees, available at http://www.nist.gov/forensics/osac/index.cfm
4. OSAC Newsletter: December 2015, available at http://www.nist.gov/forensics/osac/osac-newsletter.cfm
5. NIST and OSAC—Physics and Pattern Evidence, Scientific Area Committee/Chair: Austin Hicklin, available at https://www.nist.gov/sites/default/files/documents/2017/02/22/physicspattern_evidence_sac_intro.pdf
6. U.S. Government Publishing Office, Federal Rules of Evidence (amended January 3, 2012), available at http://www.gpo.gov/fdsys/pkg/USCODE-2011-title28/pdf/USCODE-2011-title28-app-federalru-dup2.pdf
7. Government of Canada, Canada Evidence Act (amended March 9, 2015), available at http://laws-lois.justice.gc.ca/PDF/C-5.pdf
8. Government of Canada, Weighing Evidence—Appendix A: The Rules of Evidence and the Canada Evidence Act—A3 the Best Evidence Rule (date modified August 1, 2015), available at http://www.irb-cisr.gc.ca/Eng/BoaCom/references/LegJur/Pages/EvidPreuApp.aspx
9. World Public Library, Best Evidence Rule, sourced from World Heritage Encyclopedia licensed under CC BY-SA 3.0, available at http://www.netlibrary.net/articles/best_evidence_rule
10. *Canadian Concise Dictionary*, Gage Educational Publishing Company, 2002.
11. SWGIT Best Practices for Image Authentication—Section 14 (Version 1.1 2013.01.11), available at https://www.swgit.org/pdf/Section%2014%20Best%20Practices%20for%20Image%20Authentication?docID=39
12. SWGIT Best Practices for Documenting Image Enhancement, available at https://www.swgit.org/pdf/Section%2011%20Best%20Practices%20for%20Documenting%20Image%20Enhancement?docID=37
13. SWGIT Guidelines for Image Processing (Version 2.1 2010.01.15), available at https://www.swgit.org/pdf/Section%205%20Guidelines%20for%20Image%20Processing?docID=49
14. Scientific Working Group on Imaging Technology: Documents, available at https://www.swgit.org/documents
15. Adobe Photoshop Help Files: File Formats, available at https://helpx.adobe.com/photoshop/using/file-formats.html

Color Modes and Channel Blending to Extract Detail

3

> It had long since come to my attention that people of accomplishment rarely sat back and let things happen to them. They went out and happened to things.
>
> **Leonardo da Vinci**

Back in the day of large format film cameras, we enhanced color fingerprints or prints left on colorful surfaces, ironically, with black and white film using a filter. Green filters were useful to enhance the purple tones of a fingerprint developed with ninhydrin, or an orange filter might be used to photograph a manila envelope to create contrast between the friction ridge detail and the substrate. This chapter covers the color modes in Adobe Photoshop, tools and functions available to blend channels, and provides some exercises demonstrating the multiple strategies that can be successfully applied to a single image.

Early Color Images

Scottish physicist James Clerk Maxwell lectured on the theory of creating color images by photographing a scene with black and white film through red, green, and blue filters. Transparencies of all three images would then be projected through the corresponding color filters and combined to create what would appear to be a color image. Maxwell and photographer Thomas Sutton (who actually took the photograph) created the color tartan ribbon image (Figure 3.1) in 1861 [4].

Kodak's black and white "Tech Pan" film was commonly used by forensic photographers to capture fine detail, but today, we can use color to our advantage with much more precision, control, and speed. Traditional darkroom techniques used for adjusting contrast adjustments and even sharpening are still available for improving image clarity, but digital technology offers so much more. Channel blending, contrast adjustment tools, pattern removal algorithms (fast Fourier transform or FFT), and image subtraction, just to name a few, mean that detail can be optimized in a way it never was before. Detail may be uncovered today that would have been lost to us just a few short years ago. These techniques are relatively easy to learn, requiring software and some training. Happily, many of these techniques simply require

Figure 3.1 First color image by James Clerk Maxwell and Thomas Sutton, 1861.

utilizing the tools of the trade that we have been using for decades, such as tripods, forensic light sources, and filters.

Although we will be using Adobe Photoshop to demonstrate these skills, many of these tools are not exclusive to Adobe Photoshop, and you may find other programs just as effective for color correction and image processing.

Color Models and Color Channels in Adobe Photoshop

First, a brief introduction into color is in order. There are many *color models (modes)* that have been designed to describe color. Available in Adobe Photoshop are: red, green, blue (RGB); cyan, magenta, yellow, and black (CMYK); and LAB. *Color space* refers to the various frameworks available for a color model, such as sRGB or Adobe RGB 1998.

RGB

RGB colors are *additive* colors, created by mixing spectral light in various intensities; monitors for example, create color by emitting light through red, green, and blue phosphors. The RGB color model assigns an intensity value to each pixel, ranging from 0 for black to 255 for white, for *each* color channel (red, green, and blue). When the values of all three-color components (channels) are equal, the result is a neutral tone. Computer monitors, image scanners, televisions, and most cameras are RGB devices.

Digital Color Capture

As the shutter is released on a digital camera, light reflecting from the scene comes in through the lens and is focused onto the image sensor, which is

Figure 3.2 RGB (three color channels: red, green, blue).

comprised of millions of light sensitive pixel wells, each one capable of capturing only one color, either red, green or blue. Each pixel well records the light intensity of one color only. A sophisticated algorithm then runs the data collected and interpolates that information, assigning each pixel with color and intensity values before saving it to the camera's storage medium. This is possible because of something called the Bayer array filter, a mosaic of red, green, and blue filters over the square grid of photosensors. Figure 3.2 is a representation of the Bayer array—notice that there are twice as many green filters as either red or blue. It was designed this way in an attempt to mimic human vision as closely as possible; the human eye has a great many more *green*-sensitive photoreceptors (cones) in the retina than either red or blue, making us more sensitive to green light.

The color information of an RGB image is separated into red, green, and blue channels. The contrast and detail are divided between all three channels also but not equally. More weight to detail is given to the green channel because of the extra green filters in the Bayer array (Figure 3.3).

Light intensity values for white, black, and mid-gray in RGB images:

- White = R(255), G(255), B(255)
- Black = R(0), G(0), B(0)
- Middle gray = R(128), G(128), B(128)

These three color components (channels) together reproduce all the colors you can see on screen. An RGB image contains 24 bits of color information (8 bits × 3 channels), and a 24-bit image can display up to 16.7 million colors! Considering that the human eye is capable of discerning far less in terms of

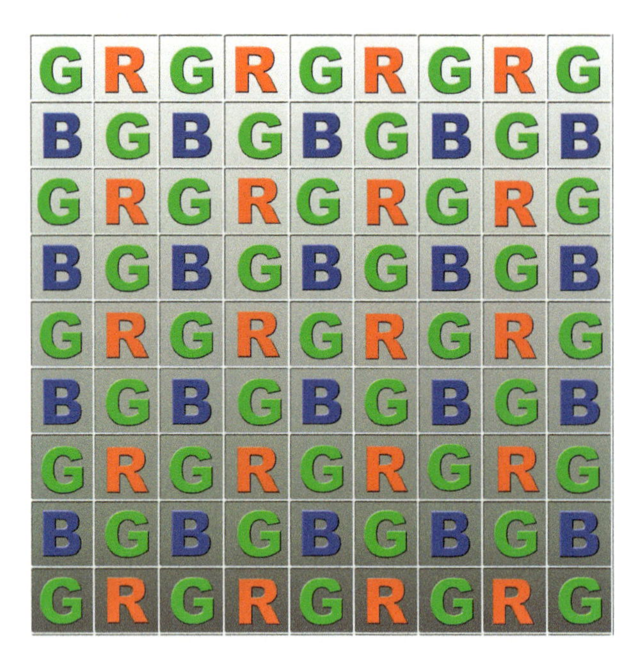

Figure 3.3 Representation of a Bayer array filter.

color distinction, the advantage is obvious; digital imaging technology offers higher sensitivity in discerning small differences in color, allowing the greatest possibility of optimizing the existing detail within an image [1] (Figure 3.4).

CMYK

CMYK colors are *subtractive* colors, and the information is separated into four channels: cyan, magenta, yellow, and black. The CMYK color model is based on the light-absorbing quality of ink printed on paper. As white light strikes the inks, some visible wavelengths are absorbed (subtracted), while others are reflected back to our eyes.

In theory, equal amounts of pure cyan, magenta, and yellow inks should result in pure black (light is absorbed—no colored light reflects back to our eyes), but because the inks are not pure, a muddy gray-brown results. Therefore, black is added as the fourth ink to represent blacks and shadows.

In Photoshop, each pixel is assigned a percentage value for each of the inks. The detail and color are distributed between four channels instead of three, and again, each one is represented as a grayscale image in the channels palette.

Ink percentage values for white, black, and mid-gray in CMYK images:

- White = C(0%), M(0%), Y(0%), B(0%)
- Black = C(80%), M(70%), Y(70%), B(70%)
- Middle gray = C(52%), M(43%), Y(43%), B(7%)

Figure 3.4 CMYK (four color channels: cyan, magenta, yellow, black).

Because there are four channels rather than three, there are many more ink combinations that can result in middle-gray than the example given above. For the purposes of this book, CMYK will refer to the default CMYK version, "U.S. Web Coated (SWOP) v2" profile.

LAB

Then there is LAB color model; it is based on the human perception of color.

There are three channels in this color model: Lightness, "a," and "b." LAB is arguably one of the most powerful and unique models offered to us in Adobe Photoshop; unlike RGB and CMYK models, the detail and contrast are kept entirely separate from the color information.

The "L" channel represents the lightness component of an image. The values in the "L" channel have a range from 0 (black) to 100 (white). Here you will find all contrast and detail—it essentially looks like a grayscale version of your image.

The "a" channel contains *only* the green and magenta color information, with values ranging from +127 (bright magenta) to −128 (bright green), with a value of 0 being neutral (neither magenta nor green). When visualizing the "a" channel of an image, anything that is lighter than 50% gray represents color that is more magenta than green. Anything darker than 50% gray represents color that is more green than magenta.

The "b" channel contains *only* the blue and yellow color information, with values ranging from +127 (bright yellow) to −128 (bright blue), with a value of 0 being neutral (neither yellow nor blue). When visualizing the "b" channel of an image, anything that is lighter than 50% gray represents color that is more yellow than blue. Anything darker than 50% gray represents color that is more blue than yellow.

For both the "a" and the "b" channels, the farther away you get from the center "0" value (neutral), the more intense that color will be. In fact, LAB color mode is capable of *describing* color so intense and bright that their actual existence is quite impossible and could never be reproduced [2]. Both channels tend to look strange and flat because there is no detail contained within them at all.

Light intensity values for white, black, and mid-gray in LAB images:

- White = L(100), a(0), b(0)
- Black = L(0), a(0), b(0)
- Middle gray = L(54), a(0), b(0)

Notice how the L number changes with the intensity of the neutral tone, but the "a" and the "b" remain at 0 because they are colorless.

Figure 3.5 shows the power of LAB when it is desirous to separate color from contrast or detail. It is obvious from the color composite that there is very little contrast in the soft magenta fingerprint, and the text is a neutral dark gray. As expected, the lightness channel displays the neutral text clearly, but the fingerprint does not show well in this channel. Fortunately, the fingerprint signal we are looking for is beautifully represented in the "a" channel, without any interfering text, because neutral contrast and detail live in the lightness channel, and only the magenta and green color data live in the "a." Even though the "a" channel is grossly lacking in contrast, it is immediately apparent that the ridge detail is clearly represented here. The "b" channel is pretty much useless, as there are not a lot distinctive yellow or blue components to the image. The bottom right corner shows the inverted version of the "a" channel, and the left corner shows the results of applying the "a" channel to itself twice to increase the contrast of this image. There is more on the Apply Image function in the following section of this chapter, but here is a brief introduction:

1. Open Figure 3.5.
2. Covert to LAB color mode. Go to Menu > Image > Mode > Lab Color.
3. Select the "a" channel from the Channels palette.
4. Convert to grayscale. Go to Menu > Image > Mode > Grayscale.
5. Invert so that the ridges are dark (Ctrl i for PC, Command i for Mac).
6. Go to Menu > Image > Apply Image.
   a. Source: The name of your image.
   b. Layer: There is only one layer in this image.
   c. Channel: There is only one channel in this grayscale image.
   d. Blending: Click the drop-down arrow and choose overlay.
   e. Opacity: 100%.
   f. Click OK.

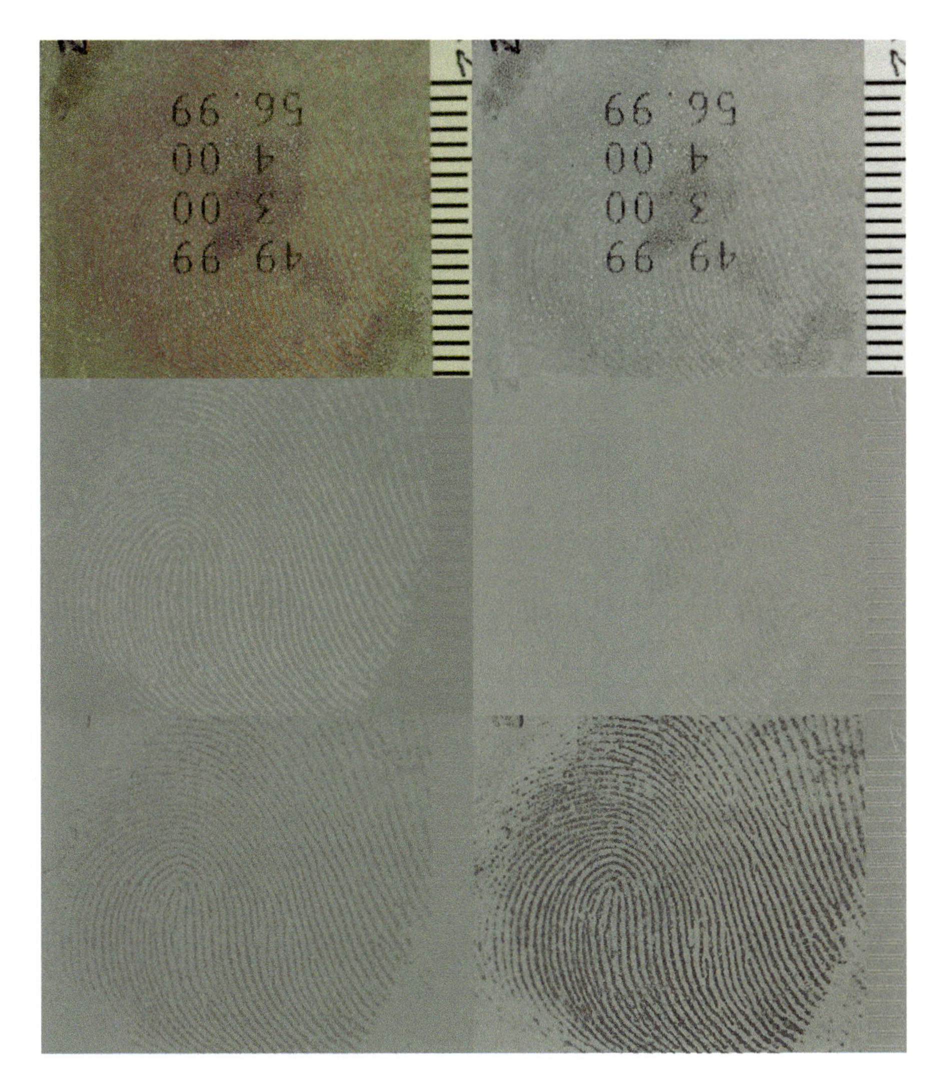

Figure 3.5 Top left to right: Composite of LAB color image, lightness channel. Middle left to right: "a" channel, "b" channel. Bottom left to right: "a" channel inverted, "a" channel applied to itself in overlay mode—twice. (Courtesy of the *Journal of Forensic Science*, Issue: 2012-5-464, The Joy of LAB Color, Figure 1. All rights reserved.)

A significant improvement is immediately apparent after blending the "a" channel with itself in overlay mode. No highlights or shadows have been clipped applying the channel to itself in this way, as there are no tonal values yet approaching pure black or white. You can apply it the same way one more time to achieve the same result that I have in Figure 3.5. This is a great starting point to begin making contrast adjustments.

Note that the Ruhemann's purple of a ninhydrin-developed fingerprint may vary in hue; sometimes the fingerprint presents a more magenta/

orangey hue, indicating a yellow component. That could mean that the fingerprint may also be visible in the "b" channel. Although experience would dictate that most of the time a ninhydrin-developed fingerprint may be found in the "a" channel, there are exceptions to every rule, so evaluating the color channels of multiple color modes and channels is a good idea.

To illustrate, consider Figure 3.6—A ninhydrin-developed fingerprint on newspaper. The substrate is noisy and distracting. Convert to LAB, and

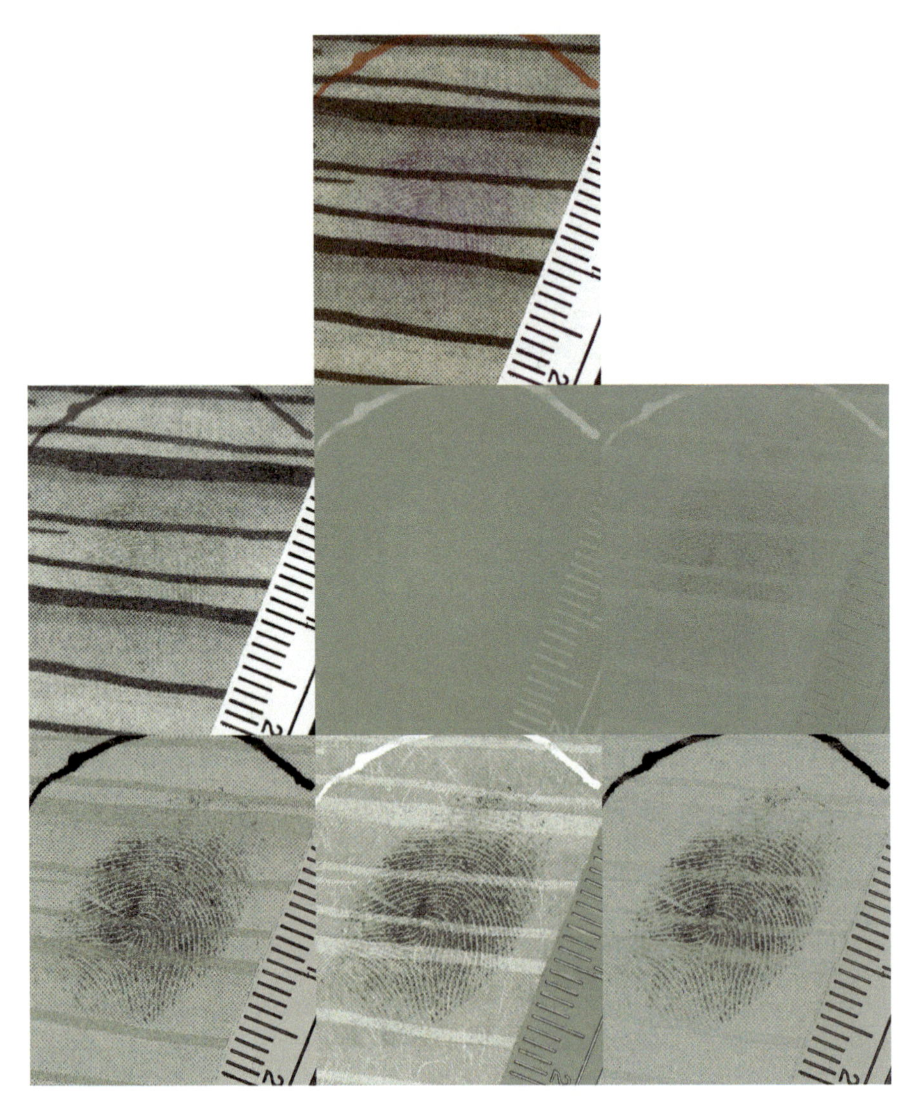

Figure 3.6 Top: Color image. Middle row: Lightness, "a," and "b" channels. Bottom row: "a" channel inverted with contrast adjustments applied, "b" channel with contrast adjustments applied; the last image is a blend of both the "a" and "b" channels (overlayed).

now there are options. The lightness channel still carries a lot of noise, but both the "a" and the "b" channels (see middle row) look somewhat promising; they both have a very compressed dynamic range, however, and require contrast adjustment. The bottom row shows an "a" channel with contrast adjustments applied (bottom left), the "b" channel with contrast adjustments applied (bottom middle), and then the final image is the result of overlaying the "a" and the "b" together (bottom right). This will be referred to as channel blending, to be addressed in the next section of this chapter.

Hopefully you begin to see the benefit of venturing into the color mode of LAB. It may not always be the answer, but sometimes the results are dramatic, revealing detail that is otherwise obscured by noise.

Grayscale

Grayscale mode is typically an 8-bit image containing one channel. Each pixel is defined as a shade of gray somewhere between 0 (black), and 255 (white), resulting in 256 shades available to describe an image (Adobe Photoshop also supports 16-bit grayscale images, which result in *over* 65,000 shades of gray). Once a channel has been isolated or channel blending techniques have been applied, resulting in a new composite channel called an alpha channel, it is advisable to convert to grayscale. This carries the benefit of faster processing because the file will be one-third of its original size. There are contrast adjustment techniques that are more effective when applied to a grayscale image, as we will learn in Chapter 6.

It should be noted that simply converting from RGB to grayscale using the color mode option in Adobe Photoshop does not blend the three channels equally. Each pixel will be converted to grayscale allotting approximately 60 percent of the green channel value, 30 percent of the red, and 10 percent of the blue [3]. Again, more weight to detail is given to the green channel because more of the green information is sampled at the capture stage with the Bayer array filter.

It is important to note the risk of losing color data when converting between color models. The simple gamut illustration depicted in Figure 3.7 shows the gamut of RGB and CMYK, with the overall shape of color being a representation of the

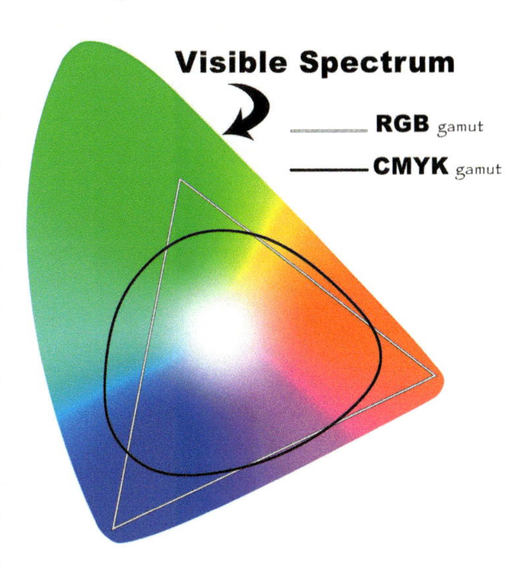

Figure 3.7 Representation of the visible spectrum, and the RGB and CMYK gamut.

colors that are visible to our human eyes [5]. Our capability to see better in the green spectrum is well represented in this illustration.

RGB has more sensitivity in the primary colors of red, green, and blue, as we would expect, yet CMYK is not capable of displaying all those hues. When converting from RGB to CMYK, any of the blues, reds, or greens that are present and described in the RGB color model but are *outside* the parameters of the CMYK model are changed to what Adobe Photoshop decides is the closest possible color. Alternatively, there are colors within the CMYK gamut that are not possible in RGB. If changes are made in CMYK, resulting in colors outside the RGB gamut, and the file is then converted back to RGB, then you will have lost some of the color data in those areas, with no control as to what those new color values will be.

For the purposes of enhancing impression evidence, you should be aware of this problem so that you do not unwittingly convert to CMYK to evaluate the channels and then convert back to RGB. You can, and should,

Figure 3.8 Compare all 10 channels of Figure 5.6. Top row: RGB image, red channel, green channel, blue channel. Middle row: Cyan channel, magenta channel, yellow channel, black channel. Bottom row: Lightness channel, "a" channel, "b" channel.

avoid this issue by *stepping back* to RGB (Ctrl > Alt > Z for PC, or Command > Option > Z for Mac) from another color model, or close and re-open your image. If you decide that you wish to use one of the channels within CMYK, no problem. Once you have blended or isolated the color channels you wish to use, the next step is to convert to grayscale directly from that channel, discarding the others (Figure 3.8).

All in all, there are 10 channels, each represented as a grayscale image, available in Adobe Photoshop to evaluate. They are all different. Ninhydrin prints are represented with the signature Ruhemann's purple. Muddy footwear may exhibit a strong yellow component. Bloody prints are red–orange. Laser and other forensic light sources may illuminate in many colors such as yellow, orange, green, and blue. Even the substrates themselves can present a myriad of colors, such as countertops, wood, tiles, and wallpaper. There are an infinite number of colorful possibilities, and the more we understand the color channels of digital images, the more we can optimize that signal detail so it can be visualized as clearly as possible. We will be evaluating fingerprints and footwear in all 10 channels, looking for one in which a strong signal could be lurking.

Photoshop Help: Channels Palette

The channels palette may be found on the right side of your Adobe Photoshop desktop. If you do not see it there, go to: Window > Channels. Figure 3.9 shows the channels palette for an image open in Photoshop. Across the top, you will see that there is a Layers tab, the Channels tab, and a Paths tab. The channels tab is highlighted because it is active. To the far right of the tab is a menu icon, the drop-down menu for channels. It offers numerous options if working in that palette. When the Layers tab is active, the drop down menu offers Layer options; when the Paths tab is active, it offers Paths options.

Underneath the tabs, there are four main bars. There is a bar each for the full color composite (LAB), lightness channel, "a" channel, and "b" channel. Each of these bars (channels) is comprised of some information:

1. Miniature eyeball—To turn this eyeball on and off means to turn the effect of the channel on and off.
2. Thumbnail—Visual representation of the channel.
3. Name of the channel (click on the *name* of the channel when selecting it).
4. Keyboard shortcut for selecting the channel—You can use this keyboard shortcut to select the channel. For example, in Figure 3.9, the keyboard shortcut Ctrl > 3 will select the lightness channel.

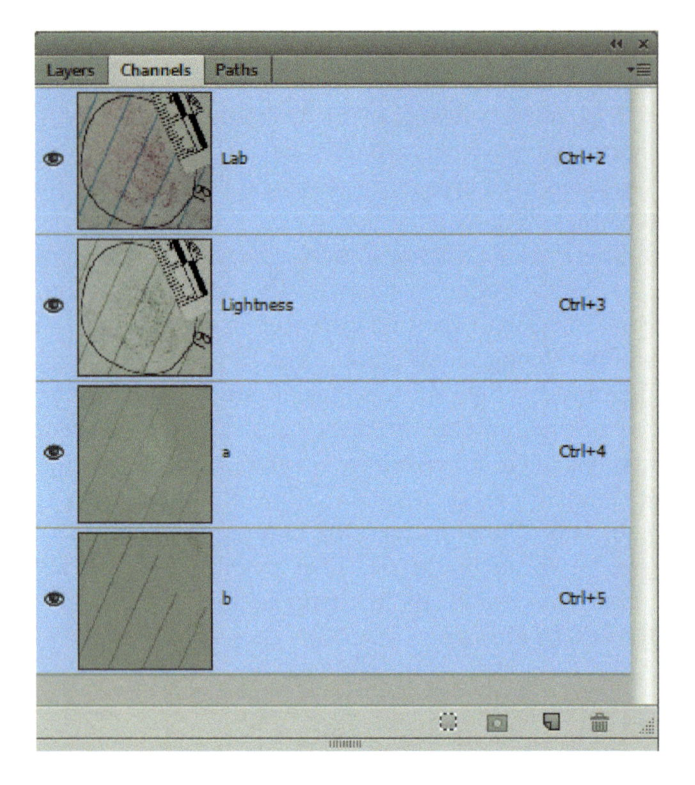

Figure 3.9 Channels palette.

Right-click the empty space under the channel bars to choose a thumbnail size (none, small, medium, large).

There are four icons at the bottom right of the channels palette. They are (from left to right):

1. Load channel as a selection
2. Save selection as a channel (like a layer mask)
3. Create a new channel
4. Delete channel

Drag either of the bottom corners of the palette to make it bigger or smaller.

Bit Depth

Bit depth refers to the number of possible intensity levels that can be displayed in an image. The higher the bit depth, the greater the number of intensity levels, referred to as tonal range. Standard grayscale and color images use an 8-bit mode tonal range. A grayscale image containing one 8-bit channel, equals 256 shades of gray that can be used to describe each pixel. An RGB image contains three channels: red, green, and blue. In a 24-bit RGB image, each of these color channels is an 8-bit grayscale representation of its color, and all 256 shades of gray are used to describe it, but there are three of them, resulting in a combined depth of 24-bit.

- $8 + 8 + 8 = 24$
- 8-bit mode—$2^8 = 256$ shades of gray
- 12-bit mode—$2^{12} = 4{,}096$ shades of gray
- 16-bit mode—$2^{16} = 65{,}536$ shades of gray

Recent versions of Photoshop now support up to 16- and 32-bit color channels in both grayscale and color images. With this increase, tonal range increases exponentially. To take advantage of the benefits of a higher bit depth during the processing stage, you would have to *capture* at the higher bit depth. Most professional grade SLR cameras are capable of shooting 12- or 14-bit images. Converting an 8-bit image to 16-bit in Photoshop does not afford any further information than was there already. With bit depth increase comes file size issues—they are going to need a lot more space.

Visually, the human eye does not decipher between 8-bit and higher bit modes. They will look the same on screen and if printed on a standard printer. The benefit of higher bit depth comes when it is time to process/edit. Instead of stretching your tonal range over 256 levels, you have many more levels to work with. This decreases the degradation of the image as contrast adjustments are being made, especially in the highlights and shadows.

Channel Blending in Adobe Photoshop

As discussed, Adobe Photoshop offers 10 channels to evaluate between the three color modes. Color information and detail has been sorted 10 different ways, and the powerful mathematical functions within Adobe Photoshop allow for blending channels. After all, every single pixel in a digital image is described with numerical values; blending channels is as easy as $2 + 2 = 4$.

Channel blending can mean anything from multiplying two channels together, adding, screening, overlaying, or more drastically, subtracting them from each other. After completing our diagnosis of the image and the channels in search of signal information, a decision can often be made to proceed with a channel blending strategy in mind.

Remember, each channel is like a grayscale image, and each pixel in it is assigned an intensity value between 0 (for black), or 255 (for white), making it a small matter to use mathematical functions between channels. For instance, when adding two channels together, the pixel value of the top left corner of one is added to the pixel value of the corresponding address in the channel to which you are adding (remember the game Battleship?). It is simple math, but wielded with a little understanding, it can be a powerful tool for clarifying signal detail that may be obscured with noise.

There are two closely related functions within Adobe Photoshop that allow us to do this:

- Apply Image
- Calculations

Both can be found under the *Menu > Image* drop-down menu.

Apply Image

The Apply Image dialogue box can be found by going to Menu bar > Image > Apply Image. In terms of recording processing steps, the Apply Image function is *not supported by adjustment layers.*

Apply Image is a function that allows one channel to be directly blended into another; the brightness and contrast results are applied directly onto the selected channel. Channels can be blended together from the same image source or from two different image sources, or a channel can even be blended with itself to increase the brightness and contrast of faint detail.

The Apply Image Dialogue Box (Figure 3.10)

Source (Image File)

The first option available in this box is the source. If there are multiple images of identical pixel dimensions open in Adobe Photoshop, they will be listed in the drop-down menu. The active image is the default selection.

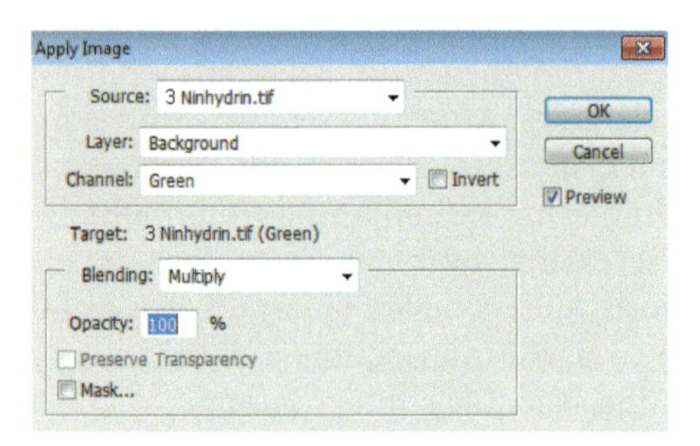

Figure 3.10 The Apply Image dialogue box (Menu: Image > Apply Image).

Layer

If an image contains multiple layers, they will be listed in the drop-down menu.

Channel

There is a choice of: Gray (a grayscale version of all the channels, though they are not equally combined for technical reasons which have been explained in the color mode section of this chapter), red, green, and blue. Beside each channel option is a check box for "invert." This allows for the flexibility to invert the channel that you are blending. This may be desirable if ridge detail is fluorescing, and black ridges are preferred.

Target

The active image name will be displayed.

Blending Mode

There are many blending modes to choose from in this drop-down menu. This book deals with the blend modes most useful to our purpose: Add, subtract, overlay, screen, multiply, lighten, and darken. These blending modes and their definitions may be found in the last section of this chapter.

Opacity

The source will be applied to the selected channel at the opacity percentage (strength) chosen. There are instances where it is not effective to apply a channel to another at 100%; these cases are visited later in this chapter.

Scale

This option is available only for the add and subtract blend modes. The result of adding or subtracting will be divided by the scale amount. The default is 1. A value of 2 in the scale amount would divide the result by 2. I recommend leaving the scale at 1 and making the brightness adjustments in offset instead.

Offset

This option is available only for the add and subtract blend modes. A positive value lightens everything equally, and a negative value darkens everything equally. For example, when adding channels, pixel values of the same address between the two are blended exactly that way—their values are added. Naturally, the result is a new channel sporting very high number values for intensity, and you may recall that only the values between 0–255 can be displayed. The offset exists to allow you to add or subtract from those values equally across the image, bringing it into a display range that can be visualized. When adding, the image might look almost entirely white initially. Setting the offset to −100 subtracts 100 equally, thereby darkening the overall result. Conversely, if a channel subtraction has been applied, the pixel values may be very low (often in the negative number range), resulting in a very low key or even mostly black image. In that case, the offset might be set to 100 to lighten the results by adding 100 to all values equally. It is important to remember to adjust the offset when adding or subtracting, as the blend may not appear as expected, looking far too dark or light, requiring the offset adjustment to make the details visible. The other blend modes incorporate an algorithm that takes care of this issue automatically; therefore, there is no offset option available in the multiply or overlay blend modes.

Chapter 3—Exercise A (Image 3.11)

Note that images used for many of the exercises in this book are available for download, so you can work along. Please visit www.crcpress.com/9781498743433. In the following example, a ninhydrin-developed print is viewed as a full color composite, as well as each of the red, green, and blue channels in turn. It is obvious that the fingerprint ridges (signal) are quite prominent in the green and blue channels. Although the blue channel does not display the ridge detail as strongly as the green channel, the blue lines (noise) running through the print are significantly muted (Figure 3.11).

There are many possible ways to approach the enhancement of the fingerprint in this image, and we will be exploring many of them, but first, to demonstrate the *Apply Image* function, we will apply the blue channel to itself to darken the ridges and see them more clearly.

Open Image 3.11. In the Channels Palette, highlight the blue channel (click on the *channel name, Blue*). Then open the Apply Image dialogue box (Menu > Image > Apply Image) (Figure 3.12).

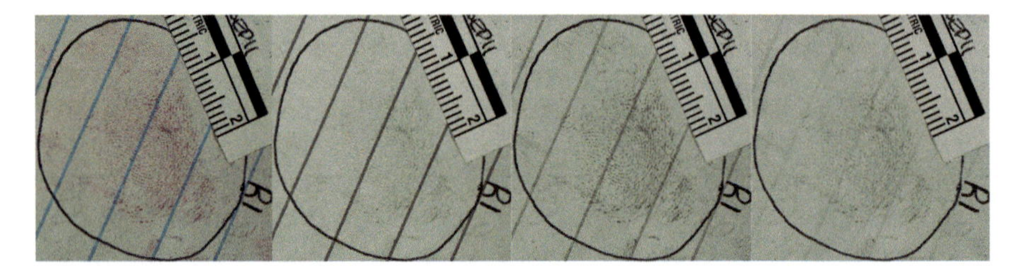

Figure 3.11 Color ninhydrin-developed fingerprint, red channel, green channel, blue channel. (Courtesy of Chris Lean, York Regional Police.)

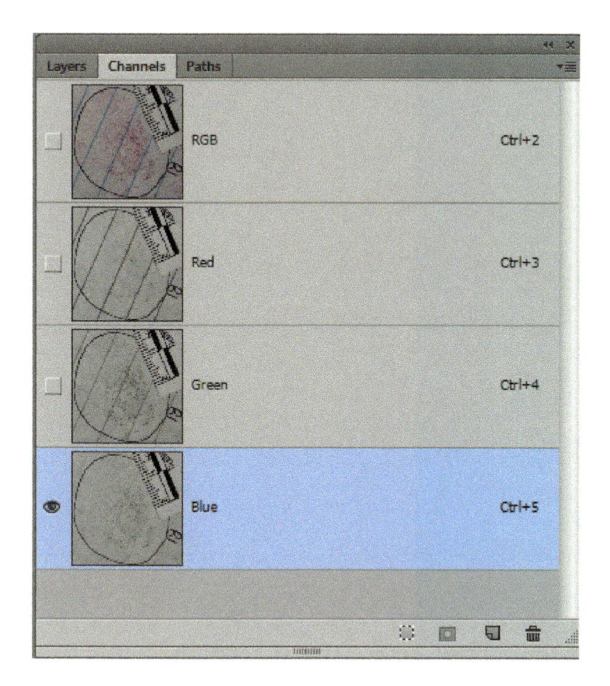

Figure 3.12 Channels palette with blue channel active.

Source (Image File)

Layer

There are no layers in this image. Leave it set to default.

Channel

Because the blue channel is selected in the channels palette, this option should default to blue. The blue channel will be blended with itself, so leave it set to blue.

Blending Mode

Compare between overlay, multiply, and add (see Figures 3.13 and 3.14). Choose overlay first.

Opacity

The blue channel in this example is blended with itself at 100% opacity.

Scale

Leave at 1.

Apply Image results: You can see that overlay lightens and increases contrast nicely, multiply mode darkens, and add increases the contrast even more. Figure 3.14a illustrates the necessity of setting an offset value when adding or subtracting (offset adjusted correctly in Figure 3.14b).

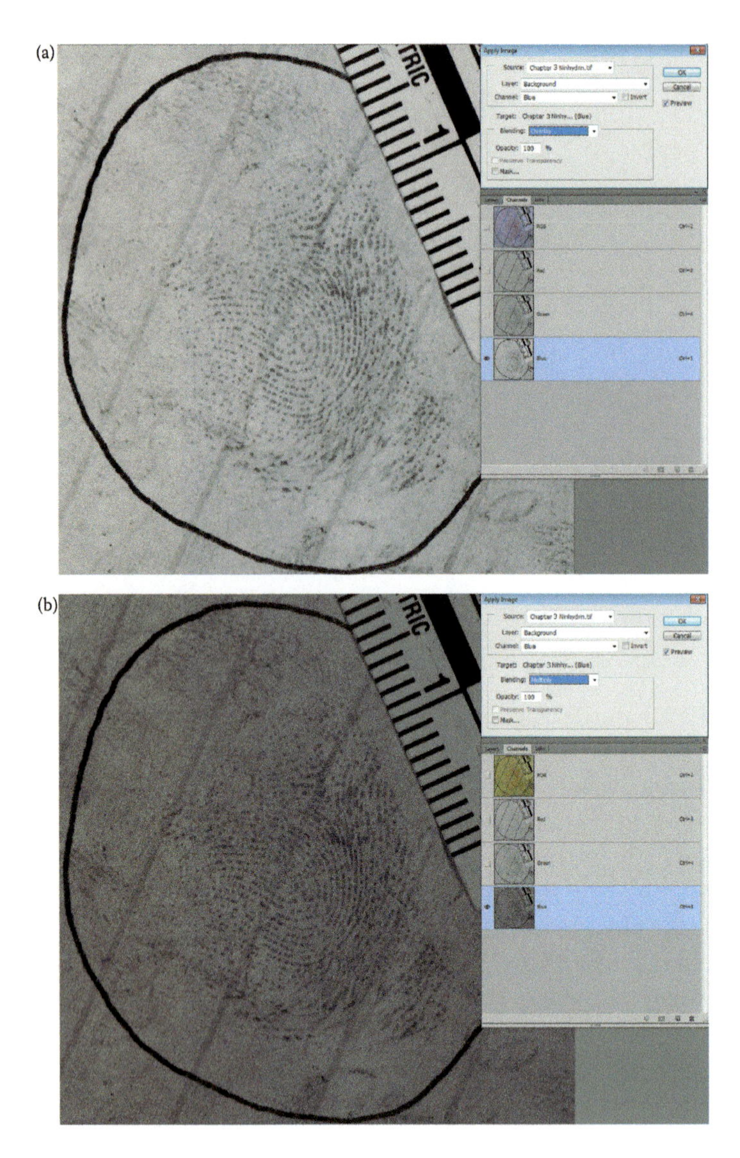

Figure 3.13 Apply Image: (a) The blue channel applied to itself in "overlay" blending mode. (b) The blue channel applied to itself in "multiply" blending mode.

This image will be used for many different blending strategies, so it becomes plainly clear that the above examples are only a few possibilities for enhancing the fingerprint ridges of this particular image. It cannot be stressed enough that there is more than one way to arrive at a destination; 10 different people may each have 10 different approaches when processing an image for analysis and still get very agreeable and similar results.

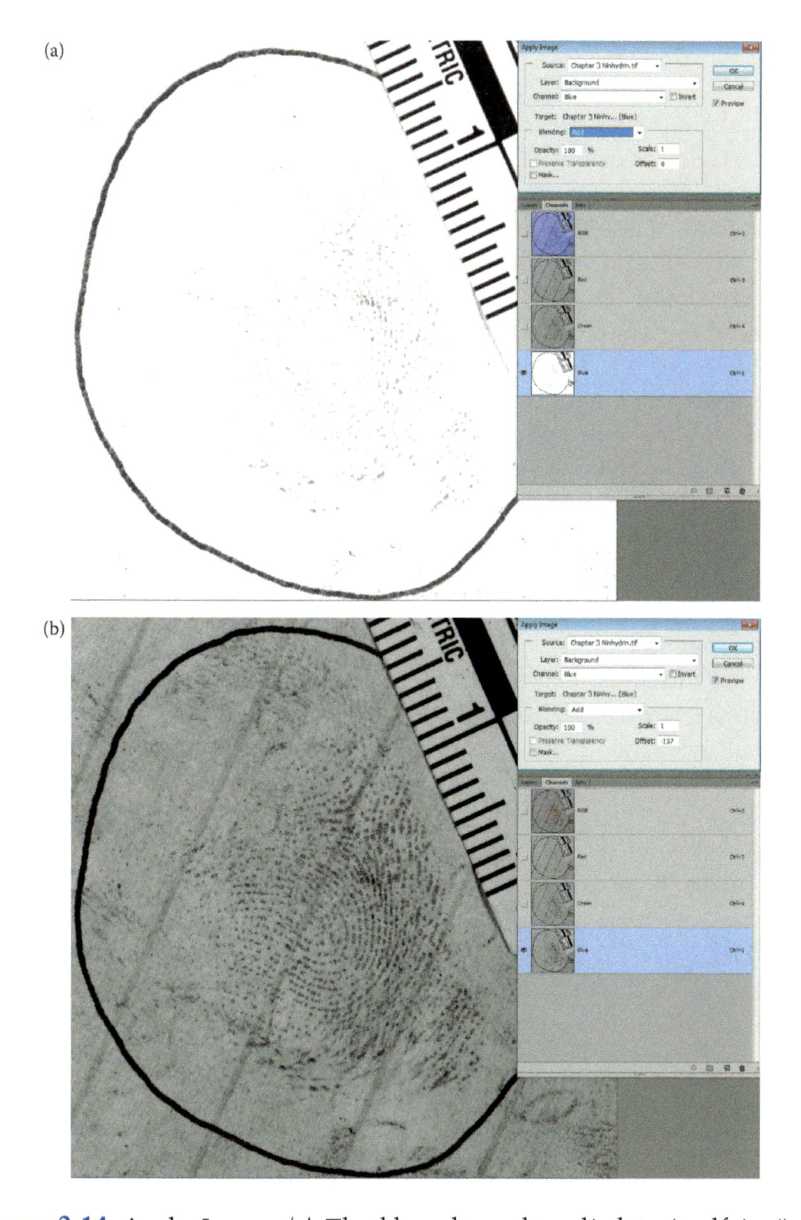

Figure 3.14 Apply Image: (a) The blue channel applied to itself in "add" blending mode. Note the offset value is left at its default of 0 to illustrate the need to adjust the offset when either adding or subtracting. It is necessary to put a negative value here when adding pixels to subtract that value from every pixel of the resulting channel. (b) The blue channel applied to itself in "add" blending mode again—this time the offset number was set to -157 (chosen visually).

Calculations

The calculations dialogue box can be found by going to the Menu bar > Image > Calculations. In terms of recording processing steps, *the calculations function is not supported by adjustment layers.*

Calculations is a close cousin to Apply Image. Two channels or images may be selected and blended together. *Both* sources must be selected in Calculations, and the result is displayed as the user chooses: either in an alpha channel in the channels palette, or as a new document (instead of being applied directly on a channel, as with Apply Image).

As with *Apply Image* functions, in *Calculations*, channels can be blended together from the same image source or from two different image sources, or a channel can be blended with itself to increase the brightness and contrast of faint detail (Figure 3.15).

The Calculations Dialogue Box

Source 1 (Image File)

If there are multiple images open in Adobe Photoshop with identical pixel dimensions, they will be listed in the drop-down menu. The active image is the default selection. It is important to remember that Source 1 is applied *to* Source 2 in the blending manner selected.

Layer
If an image contains multiple layers, then select one from the drop-down menu.
Channel
Select a channel for source one from this drop-down menu.

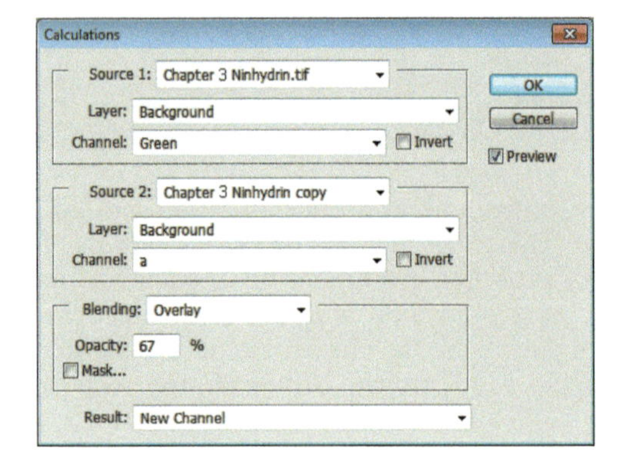

Figure 3.15 Calculations dialogue box.

Source 2 (Image File)

The second source does not have to be the same as Source 1. It is possible to blend channels between different source images, as long as their pixel dimensions are identical.

Layer
If an image contains multiple layers, they will be listed in the drop-down menu.
Channel
Select a channel for source two from this drop-down menu.

Blending Mode

Select the desired mode for blending sources together.

Opacity

Source 1 is applied *to* Source 2 using the blending mode chosen, at the opacity percentage set here. There will be instances where it is appropriate to blend Source 1 *to* Source 2 at a reduced opacity.

Scale

Leave this value at 1.

Offset

Offset value must be set when using blending modes Add and Subtract; a positive value lightens the result, while a negative value darkens the result.

Result

The applicable options for our purposes is to either leave it set to new channel and allow the result to be displayed in an alpha channel, or to click the drop-down arrow and choose new document, whereupon the result is opened in a new and untitled document. The next step in the workflow is often to convert to grayscale anyway, after completing the color blending process, so this is a matter of personal preference.

Chapter 3—Exercise B (Image 3.11)

Open the same image again in Photoshop (3.11). Duplicate the image: Menu > Image > Duplicate Image. You are prompted to give this extra copy a name—ignore that and click OK. There are now two identical RGB images on the desktop. Convert one of them to LAB color mode (Menu > Image > Mode > LAB color).

Review the channels in the channels palette. The lightness channel looks like a grayscale version of our image, and it is not bad; it certainly could be used to enhance the ridges. At first blush, the "a" and "b" channels look useless; the severely compressed dynamic range leads us to believe that there is little of value, but look a little more closely at the "a" channel, and you can see that the ridges are very light (anything lighter than 50% gray is more magenta than green), and the blue lines in the paper are dark. To invert the "a" channel, go to Image > Adjust > Invert (Ctrl i for PC, Command i for Mac), to make the ridges dark. The blue lines in the paper are now *lighter* than the background, which is interesting because in the green channel the ridges are dark, and the blue lines were also dark. What happens if we blend the green channel (RGB) and the "a" channel (LAB) together? Blending the dark ridges of both channels together to make them visibly stronger helps the fingerprint signal, but what happens to the blue lines in the paper? These lines are dark in the green channel and light in the "a" channel. The idea is to cancel out the noise while strengthening the fingerprint signal simultaneously.

Before blending the inverted "a" channel with the green channel, it needs some contrast adjustment. With the "a" channel selected, apply it to itself in overlay mode. Go to Image > Apply Image. The LAB image is active on the desktop, so the source should default to the LAB image. The "a" channel is selected in the channels palette, so the channel option should default to "a." Choose a blending mode of overlay. Click OK. This is a great way to optimize the low contrast channels of "a" and "b." A channel can be applied to itself more than once, as shown in Figure 3.16 (bottom right image—the "a" channel inverted and applied to itself in overlay mode twice).

Blending the green channel from RGB and the "a" channel from LAB, open the calculations dialogue box: Image > Calculations (Figure 3.17).

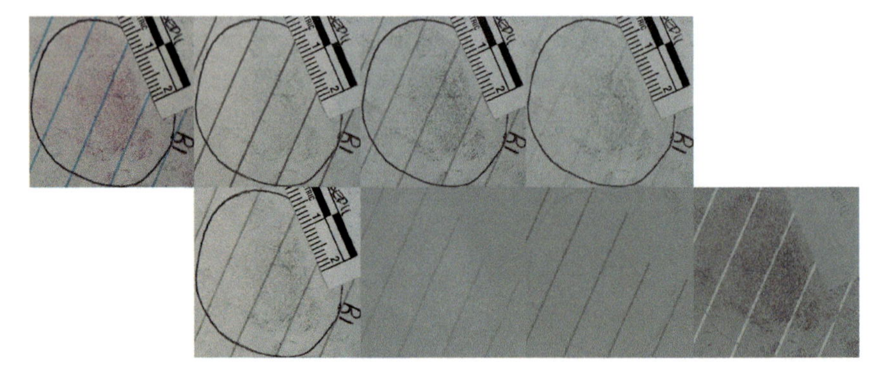

Figure 3.16 A review of image channels in RGB and LAB: Top row left to right: Color composite, red channel, green channel, blue channel. Bottom row left to right: Lightness channel, "a" channel, "b" channel, and the "a" channel after inverting it and overlaying it with itself twice.

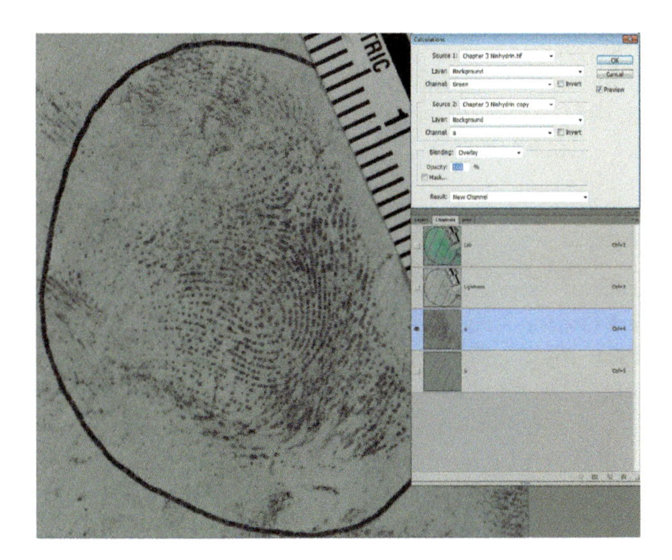

Figure 3.17 Calculations overlaying the green channel of RGB at 100% strength with the "a" channel of LAB. The result is a new alpha channel in the channels palette. The blue lines in the paper identified as "noise" have been significantly muted.

- Source 1: Click the drop-down arrow and choose the RGB image
- Source 1 Channel: Green
- Source 2: Choose the LAB image
- Source 2 Channel: "a" (it has already been inverted to make ridges dark)
- Blending mode: Overlay
- Opacity: 100% (or choose your own percentage for blending the green into the "a")
- Result: New channel
- Click OK

There is now another channel in the channels palette called *alpha channel*. This is one of the main differences between the *Apply Image* and *Calculations* functions, as *Apply Image* applies your changes directly to the active color channel, and *Calculations* makes no adjustments in the channels at all but creates the blend in a brand new one. This may be beneficial when it is desirable to blend parts of the original channels to the alpha channel being built.

Try the exercise again with multiply—a darker result, but with the contrast tools at our disposal, a good result can be achieved from either of the blend modes.

The result of this calculations blend is then converted to grayscale (Image > Mode > Grayscale. Photoshop will ask if you wish to discard the other channels—Click OK) and adjusted using one or any combination of contrast adjustment tools before it is saved as a processed image (levels, curves, shadows and highlights, noise removal, unsharp mask, etc.), with some form of notes to track the processing steps taken to achieve the end result.

Chapter 3—Exercise C (Image 3.11)

Not to be ignored, CMYK offers many benefits to color fingerprint work as well. Open Figure 3.11 again, and convert to CMYK (Image > Mode > CMYK color. Photoshop will flag you that you are about to convert to CMYK using the "U.S. Web Coated (SWOP) v2" profile. Click OK). The fingerprint detail can be visualized clearly in the magenta channel. At this point, you would not be amiss to simply convert the magenta channel to grayscale and make contrast adjustments, but hold on a minute—the cyan channel is sporting some seriously strong noise. There is some faint indication of ridge detail, but overall the noise is dark and overpowering. What if we attempted to *subtract* the cyan channel *from* the magenta channel? Because the noise is so strong in the cyan, we know it is not necessary to subtract it from the magenta at 100% strength, as that will just create noise in the opposite direction, manifesting itself as light lines across the image, rather than dark; instead, we can accomplish the subtraction at a smaller percentage. Yes, it's very cool (Figure 3.18).

Open Figure 3.11. Go to Image > Calculations. Remember that Source 1 is applied *to* Source 2 using the blending mode you select. Set Source 1 to cyan channel to subtract *from* magenta channel in Source 2—opacity around 37% (adjust this percentage value up or down). The offset value for this example is set at +135 (adds 135 tonal values to every pixel, lightening the result Figure 3.19).

The point of analyzing the same photo in so many ways is simply to illustrate that there are so many possibilities. You think this image is convenient for these examples but unlikely in real casework? Think again. They all won't be convenient like this, sure, but they *do* happen frequently. It's a beautiful thing.

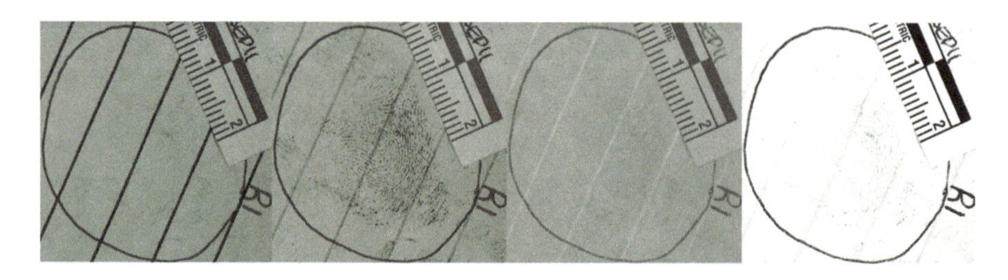

Figure 3.18 The CMYK channels from left to right: Cyan, magenta, yellow, and black. The cyan channel sporting a very strong noise pattern, and the magenta channel has some noise and fingerprint signal.

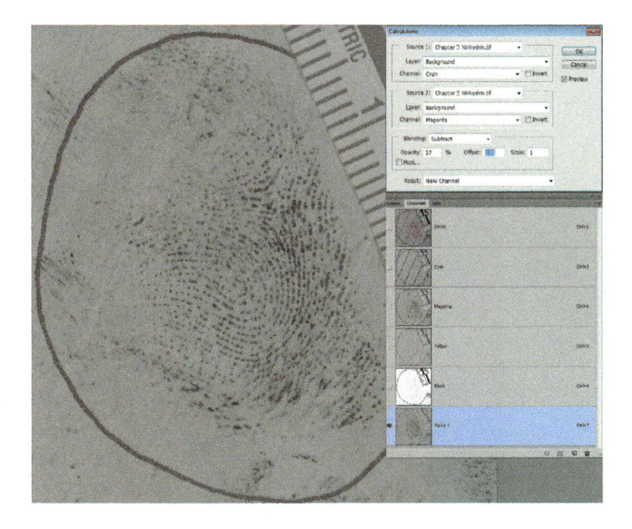

Figure 3.19 The Calculations dialogue box and the results of subtracting the cyan channel from the red channel at 37% opacity and offset at +135. The channels palette shows the new alpha channel created as a result.

Chapter 3—Exercise D (Image 3.20)

Figure 3.20 is already a great print and needs very little in the way of processing, but it is a great example of how simple the processing steps can be and how many solutions there can be to an imaging problem. First, review the color channels. Figure 3.21 shows an overview of the RGB, CMYK, and LAB channels.

Figure 3.20 Color image—Fingerprint developed over blue ink writing. (Courtesy of York Regional Police. All rights reserved.)

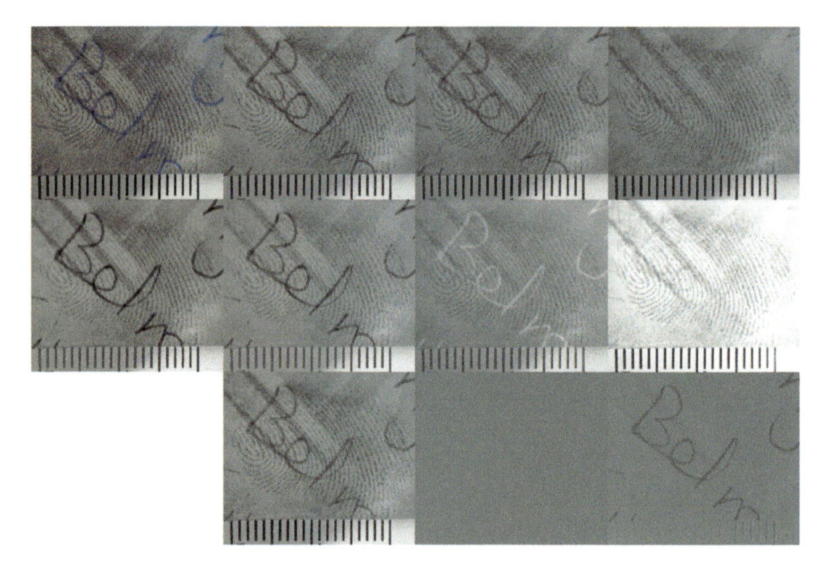

Figure 3.21 Top row left to right: Color image, red channel, green channel, blue channel. Middle row left to right: Cyan channel, magenta channel, yellow channel, black channel. Bottom row left to right: Lightness channel, "a" channel, "b" channel.

Four Ways from Sunday

1. The blue channel immediately strikes me as the most logical choice. In RGB color mode, select the blue channel, convert to grayscale (Image > Mode > Grayscale), and the image is ready for some simple contrast adjustment. That was easy.
2. A variation of the first option: Apply the blue channel to itself in overlay mode. Select the blue channel (remember when selecting channels, to click on the *name* of the channel; *do not* click the eyeball, or things will go awry). Go to Image > Apply Image, ensuring that channel is set to blue, and blending mode is set to overlay. Opacity 100%. Convert to grayscale (Image > Mode > Grayscale) (Figure 3.22).
3. One of the students in our workshop converted the image to CMYK (Image > Mode > CMYK color), then blended the cyan channel where the blue ink is represented darkly, with the yellow channel, where the blue ink is represented as a light tone (refer to Figure 3.20). Open the calculations dialogue box (Image > Calculations), and set Source 1 channel to cyan, Source 2 channel to yellow, the blend mode to overlay, and the opacity may be anywhere between 80 and 100% (Figure 3.23).
4. Another student had a different idea; they saw that the "b" channel in LAB color mode was almost entirely noise with no fingerprint detail at all, and the lightness channel had good fingerprint detail with some of the noise present. They decided to subtract that noise

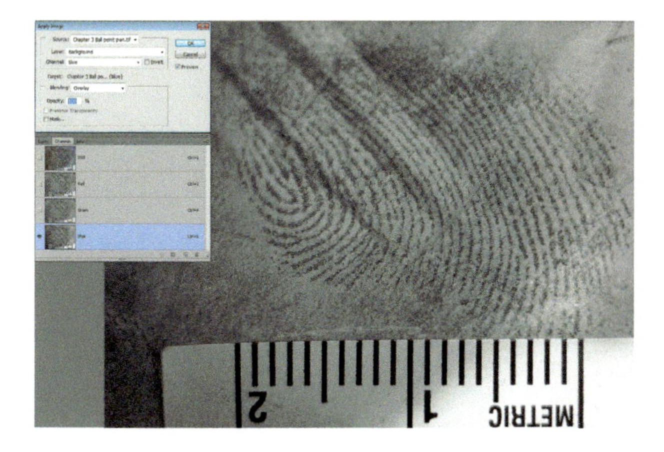

Figure 3.22 Image of Figure 3.20—(RGB) The blue channel applied to itself in overlay mode.

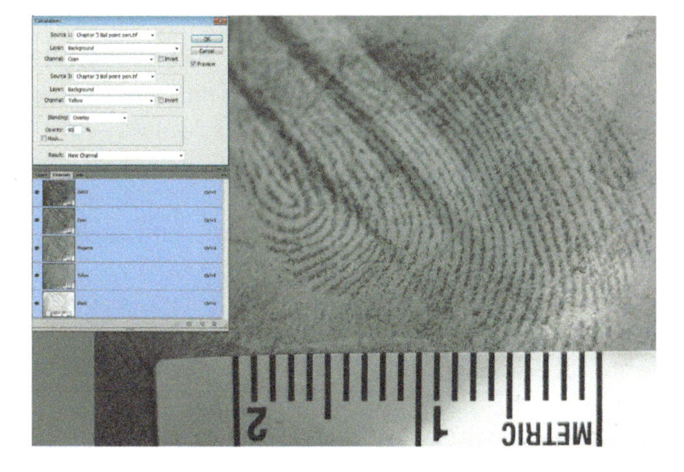

Figure 3.23 Image of Figure 3.20—(CMYK) The cyan channel blended with the yellow channel in overlay mode at 90% opacity.

from the lightness channel. Convert the image to LAB (Image > Mode > LAB color), enhance the "b" channel (the noise) by applying it to itself in overlay mode twice, *then* open the Calculations dialogue box (Figure 3.24):

a. Source 1—Channel: "b"
b. Source 2—Channel: L
c. Blending: Subtract
d. Opacity: Around 70%
e. Offset: Around 125

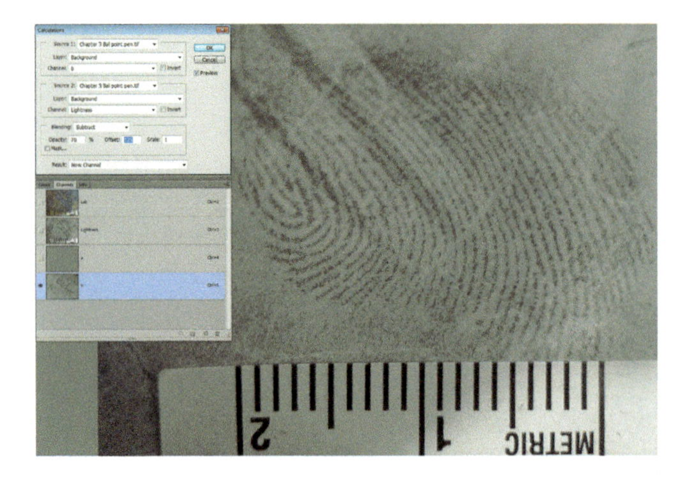

Figure 3.24 Image of Figure 3.20—(LAB) The "b" channel enhanced (overlay "b" with itself), subtracted from the L (lightness) channel at 70% opacity. Offset 125.

There are even more possibilities than what has been covered here, and that's one of the reasons this image is so great to work while exploring channel blending techniques. The students in options 3 and 4 opted for a more advanced approach to enhancing this image, ending up with great results—I like the way they are thinking! Sometimes it is as simple as choosing a great channel to work with, converting to grayscale, and applying a contrast adjustment.

Chapter 3—Exercise E (Image 3.25)

Figure 3.25 is a fingerprint treated with Ardrox on a AA battery. The photographer asked if anything could be done to clarify the fingerprint for analysis as the ridge detail was obscured by text. I was doubtful and wondered if Rhodamine 6G or Brilliant-Yellow treatment would have induced a stronger fingerprint signal, thereby setting it further apart from the substrate. However, I assured him I would explore the image to see if there was anything that could be done to help it, and to my surprise, the substrate was easier to subdue than I had anticipated (Figure 3.26).

Open Figure 3.25. The RGB image itself is sufficient to accomplish a good result; the fluorescing fingerprint is best represented in the green channel, and you will notice the blue channel seems to be a perfect representation of the noise that is distracting from the fingerprint detail. The plan: subtract that noise from the green channel. Go to *Calculations*. Set the Source 1 channel to blue, and *check the invert box beside it*; set the Source 2 channel to

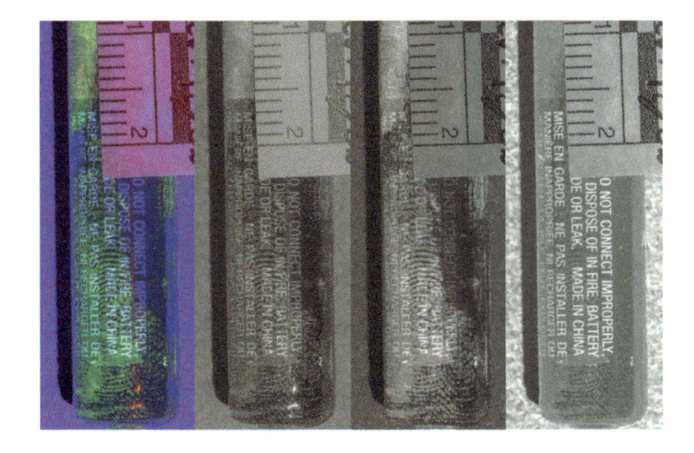

Figure 3.25 Left to right: Original color composite, red channel, green channel, and blue channel. (Courtesy of Greg McGuire, York Regional Police. All Rights Reserved.)

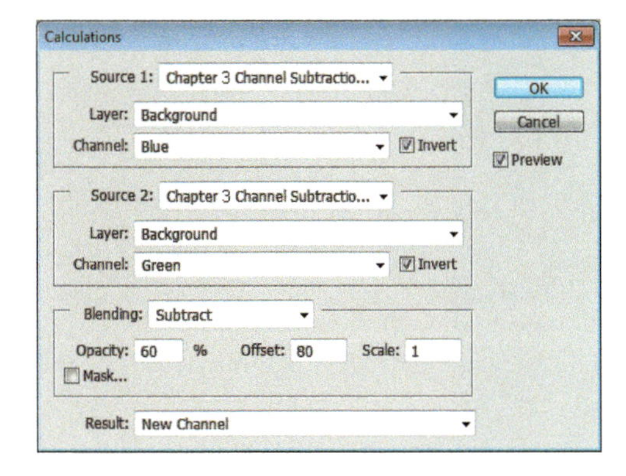

Figure 3.26 Calculations dialogue box. Blue channel (inverted) is subtracted from the green channel (inverted) at 60% opacity and an offset value of 80.

green, and *check the invert box beside it also* (inverting it will make the ridges dark in the results—they are fluorescing in the original). The blending mode is subtract, but the noise in the blue channel is somewhat overwhelming for this image. Reduce the opacity of the blue channel until the noise starts to blend into the background. Do not forget to adjust your offset to a value—contrast will be adjusted later), click OK.

Figure 3.27 left shows the results of the subtraction. To the right is the image after applying shadow/highlight and levels (see Chapter 6 for contrast adjustment tools).

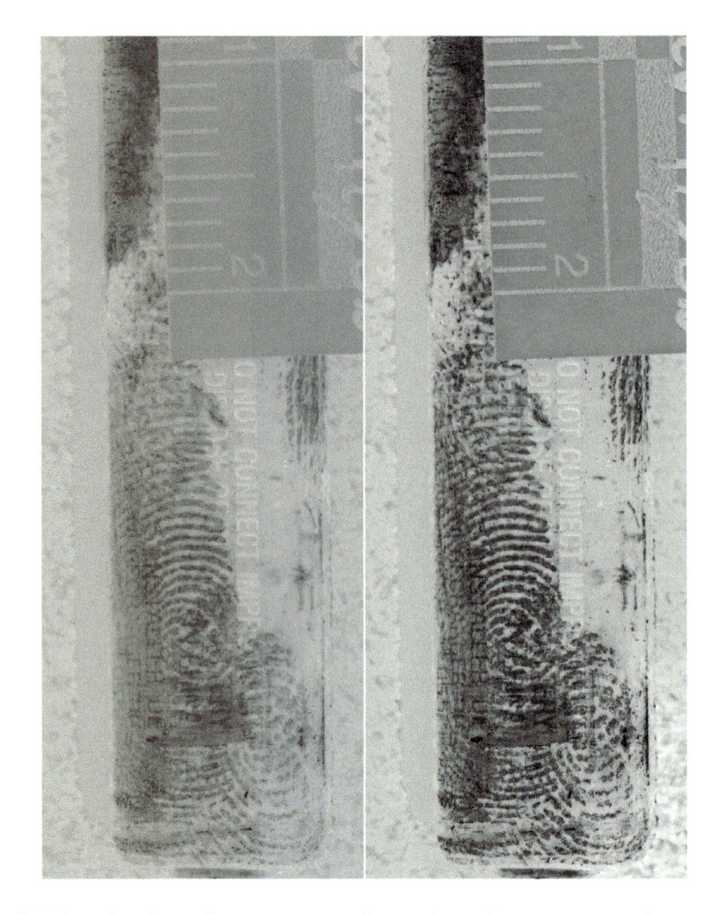

Figure 3.27 Left: The subtraction result. Right: After contrast adjustments.

Channel Mixer

The channel mixer dialogue box can be found by going to the Menu bar > Image > Adjustments > Channel Mixer. *It may also be added as an adjustment layer to your image file for tracking processing steps.*

The channel mixer is another way of blending color channels within the same image using color channel sliders. A slider for each color channel in the image allows you to add that channel to your monochrome result in percentages. A real-time preview allows visualization of the results as you mix color channels (Figure 3.28).

1. Output channel—Once the monochrome box has been checked (bottom left of box), there is only a gray option.
2. Source channels—There are sliders for mixing for each color channel in the image (red, green, and blue for RGB and cyan, magenta,

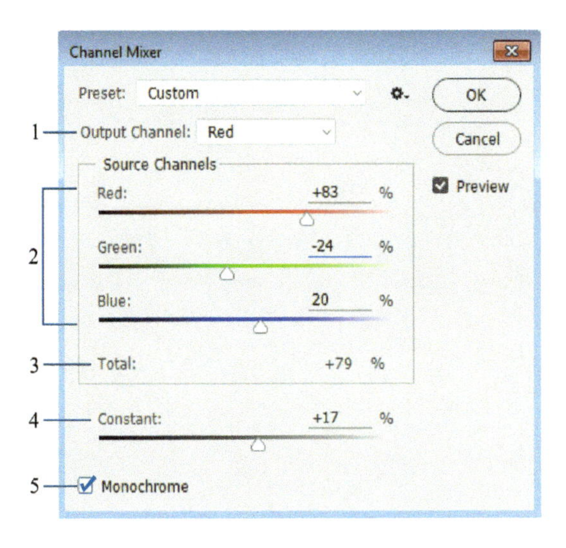

Figure 3.28 (1) Output Channel. (2) Source Channels. (3) Total. (4) Constant. (5) Monochrome.

yellow, and black for CMYK). This function is not available with LAB color mode.

3. Total—The total percentage of channels mixed is displayed under the sliders. You are increasing or decreasing the presence of any color channel by a percentage as you drag its slider. It is advisable to keep this total percentage under 100%. Anything over 100% will be flagged with an exclamation mark to warn you against clipping image values.
4. Constant—Drag this slider to lighten or darken the image.
5. Monochrome—Check this box before beginning your adjustments to see the effects in grayscale. If you have added channel mixer as an adjustment layer, the monochrome check box may be found at the top of the dialogue box.

Figure 3.29 features red print that is detracting from the fingerprint detail. Whether you are opening channel mixer from the menu bar or as an adjustment layer, the first step is to check the monochrome box. Increasing the percentage of the red channel in the mixer decreases the red ink contrast. To see the effects that each channel has on an image, increase the first one to 100% while leaving the others at 0%. Repeat this exercise for each channel slider. This gives you a good idea where you should start with the siders, and then you can fine tune your adjustments.

When done, covert to grayscale (unless using adjustment layers), and evaluate the image to determine if further contrast adjustments are required.

Figure 3.29 Settings applied to blend channels in channel mixer.

Black and White

The black and white dialogue box can be found by going to the Menu bar > Image > Adjustments > Black and White (PC: Alt, Shift, Ctrl B, Mac: Option, Shift, Command B). It *may be added as an adjustment layer to your image file* to track processing steps, if that is your recording method of choice.

The black and white function offers six color sliders for all of the primary and secondary color channels, and it is good for creating grayscale images from color where signal information may be difficult to see due to color interference (see Figure 3.30).

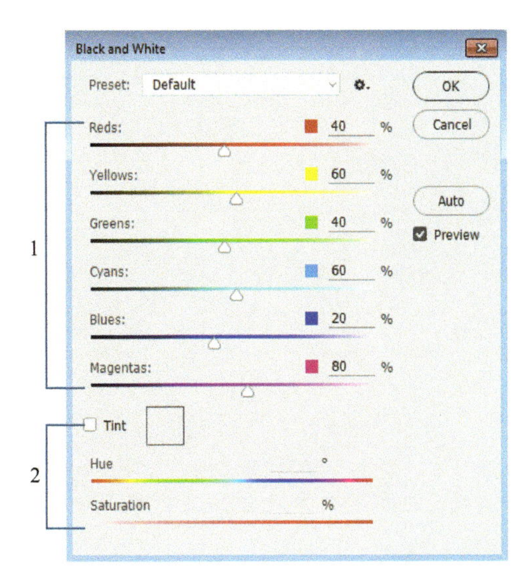

Figure 3.30 Black and White dialogue box. (1) Color sliders. (2) Tint options.

1. Color sliders—As with the channel mixer, the primary color sliders are available for adjustment; unlike channel mixer, the secondary sliders are also available for adjustment.
2. Tint options—This section of the dialogue box deals with tinting an image with color and saturation adjustments. There is no forensic function to this.

There is no monochromatic check box in the black and white dialogue box; it automatically creates a grayscale preview. This function also allows you to click and drag in the image to modify a slider. Move your cursor over an area you wish to adjust; click and drag. As you drag right and left, the brightness for that color is adjusted.

Note for users recording steps with adjustment layers: The *"click and drag in image to modify a slider"* button must first be activated for this click and drag feature to work. It is near the top of the dialogue box, and the icon looks like a pointing hand with double arrows shooting right and left.

Figure 3.31 shows the before and after images with the Black and White settings used to subdue the red printing on the substrate. When done, if you are not using adjustment layers, convert to grayscale after making adjustments, and proceed to contrast adjustment tools.

While many of these blending techniques may seem strange and foreign, do not be intimidated. The goal is to find a channel in which the signal presents itself most strongly with the least interference of noise. If you are working with a good quality fingerprint image, and it doesn't need any further processing to analyze with ease, then don't do any! Some of our examples in

Figure 3.31 Settings applied to adjust colors with black and white.

this book are good fingerprint images already; they were chosen because they best illustrate a concept or idea.

In closing this chapter, the best advice we can give you is this: Understand what you are doing and why. Don't press yourself beyond your level of understanding; it is *you* who must explain what you did and why you did it. Make sure you are comfortable with being able to do so with the images you process, and you can't go wrong.

Blending Mode Definitions (Channels)

Multiply

Darkens—The algorithm multiplies the values for each pixel in both channels specified and then divides that by 255 (always a darker result). Multiply may be useful for blending channels with faint ridge detail in ninhydrin, for instance.

Screen

Lightens—The algorithm multiplies the inverse values for each pixel in both channels specified, resulting in lighter values.

Overlay

Contrast boost—Multiplies dark pixel values and screens light pixel values of the base image/channel; dark areas become darker, and light areas become lighter. Blending with 50% gray has no affect.

Add

High contrast—The pixel values of both channels are added together. Any resulting values higher than 255 (white), are clipped to 255; therefore, the offset (explained in the following example) must be adjusted to darken it within the 0-255 visual range.

Subtract

The values of the blend channel (Source 1) are subtracted from the base channel (Source 2). Any resulting negative values are clipped to 0 (black); therefore, the offset must be adjusted to lighten it within the 0–255 visual range.

Lighten

The values of the blend channel are compared to those of the base channel. The lighter of the two becomes the resulting value for that pixel.

Darken

The values of the blend channel are compared to those of the base channel. The darker of the two becomes the resulting value for that pixel.

Review Questions

1. Name three color modes available in Adobe Photoshop.
2. Which color mode is particularly unique and why?
3. There are many ways to blend color channels together. Name four functions or tools within Photoshop that can be used to do this.
4. Which channel blending function allows the blending of two channels between two different images?

References

1. Number of Colors Distinguishable by the Human Eye, available at https://hypertextbook.com/facts/2006/JenniferLeong.shtml
2. D. Margulis, *Photoshop LAB Color: The Canyon Conundrum and Other Adventures in the Most Powerful Colorspace*. California: Peachpit Press, 2006.
3. D. Margulis, *Professional Photoshop: The Classic Guide to Color Correction*, Fourth Edition, New York: Wiley Publishing, 2002.
4. National Geographic: Milestones in Photography, available at http://photography.nationalgeographic.com/photography/photos/milestones-photography.html#/niepce-first-photo_1459_600×450.jpg
5. Cie Color Gamut Illustration, available at http://dba.med.sc.edu/price/irf/Adobe_tg/models/rgbcmy.html

Multiple Image Techniques

4

It's fine to celebrate success, but it is more important to heed the lessons of failure.

Bill Gates

The purpose of this chapter is to illustrate methods in which capturing more than one image of an impression (fingerprint, shoeprint) can provide greater opportunity for improving the clarity and strength of that image. Today, some forensic practitioners conduct their activities exclusively in front of a computer monitor, while others with a more generalist job description are responsible for impression photography as well. In either case, optimum quality images expand the possibilities for success in image processing.

Fingerprints, footwear, and other impressions can be characterized as signal. The development and detection of this evidence routinely results in some degree of interference, consisting of color, repetitive pattern, or random anomalies, which constitute noise. The photographer's goal in *every* evidence image is to *optimize the signal-to-noise ratio.*

Identification photographers of the film era did not have the luxury of instant gratification. They could not examine their images at a homicide scene immediately after capture to confirm that they were properly focused, exposed, and illuminated. The most notable example of multiple image capture in that period was bracketing, taking photos above and below the metered exposure, to ensure that in difficult or challenging lighting situations at least one of the negatives recorded all the subject detail. The results would not be revealed until the film had been developed and printed and the individual negatives examined, sometimes hours or even days after the photographs had been taken.

All the photographer's art and skill, in terms of optimizing the subject and minimizing substrate interference (signal to-noise ratio), had to be applied in the preparation stage prior to tripping the shutter—aperture selection, lighting, exposure, and focus. The film was then subjected to chemical processing in a darkroom. As soon as the film development stage had been completed, the negative was stabilized and, in the context of conventional photography, could not be changed. There were limited options in the printing process for making relatively minor alterations to the contrast and the dynamic range of the final print. These included paper choice (contrast grade) and dodging and burning.

Realistically, if a fingerprint recorded on a film negative was obstructed by background interference (color, pattern) and lay just beyond the threshold of suitability for comparison, that was very likely where it was going to remain.

In the realm of digital imaging, so much more is possible in terms of optimizing signal-to-noise ratio. Images can be combined or blended in several ways, including the four basic functions in arithmetic—addition, subtraction, multiplication, and division. Identification practitioners have two sizeable advantages in the digital domain that did not exist in the world of film. First, there is significant potential for revealing more detail in images of fingerprints and footwear in post-photography processing. In most cases, one well-taken image of a fingerprint or footwear impression is all that's required for optimization in Photoshop or other similar software program.

Second, and the focus of this chapter, challenging situations of background obstruction can be diagnosed prior to taking the photograph and, in some cases, even before the chemical processing decision has been made. A multiple-exposure technique can often provide substantially greater opportunities for increasing the signal-to-noise ratio. Succinctly stated, it can make the difference between a tantalizing fragment of ridge detail that remains just out of reach and an impression suitable for comparison.

Image Subtraction

Image subtraction was first explored in the field of identification in the 1990s and early part of this century [1–3], a time when digital images were routinely captured on a grayscale tube camera interfaced with an analog/digital converter. Evidence images of that period were small in size (500 KB) compared with current resolution. Each pixel of the image had a grayscale value between 0 (black) and 255 (white).

Two images (or two channels) are required for a subtraction. First, an image containing both signal (fingerprint, scale, and markings) and interfering substrate is acquired. Ideally, the goal is to capture a second image containing only the substrate, with the fingerprint, scale, and markings absent. Photoshop can be used to subtract one image from the other, pixel by pixel. Elements common to both images will subtract themselves out of existence. The substrate, appearing in both images, will subtract itself and be greatly diminished or absent in the resulting image, leaving the desired items obstructed to a lesser degree (Figure 4.1).

Wherever there has been no change, the pixel in the resulting subtracted image has a value of zero (0) or black. This can be displayed quickly and easily.

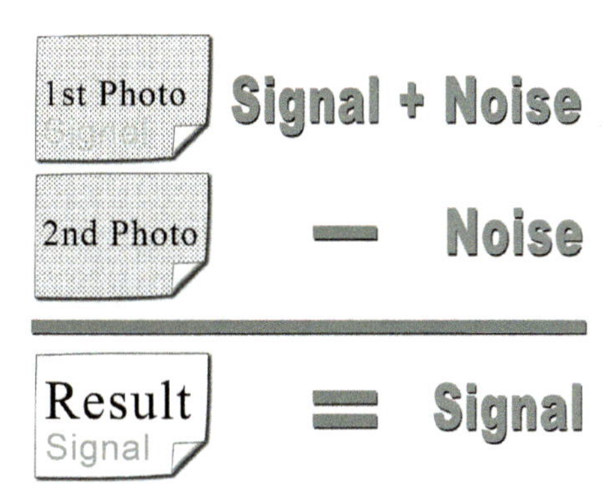

Figure 4.1 When you subtract noise from signal and noise, the result is just signal.

- Open any image in Photoshop.
- Select *Image* in the top row of options.
- Select *Calculations* near the bottom of the drop-down menu.
- The Calculations window will appear, displaying the active, highlighted image as both Source 1 and Source 2.
- Immediately below both source windows are *Layer* and *Channel* spaces.
- Next is the *Blending* window, which will display Multiply, when Calculations is first opened (Figure 4.2).
- Using the drop-down arrow in *Blending*, select *Subtract* (Figure 4.3).
- Ignore (for this operation) *Opacity* and *Scale*.
- Below the *Blending* option, the *Offset* window can be seen.
- Set the *Offset* value at 0.
- Immediately, the live image window will be uniform black, showing the result of subtracting one image from itself.
- The image resulting from a subtraction is not a true image. It is composed of the grayscale *differences* at each pixel between the two

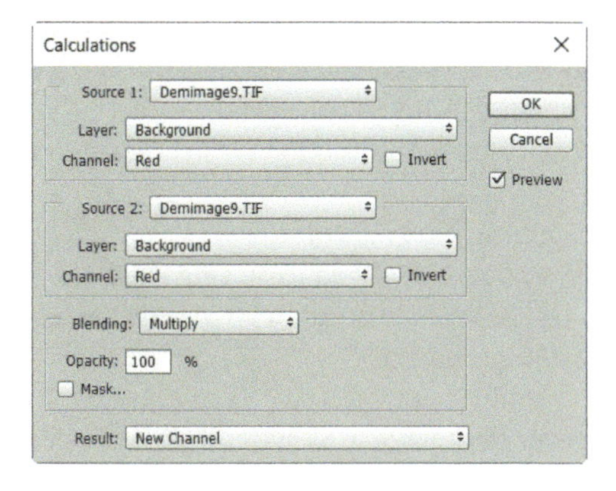

Figure 4.2 Calculations window, showing Blending option.

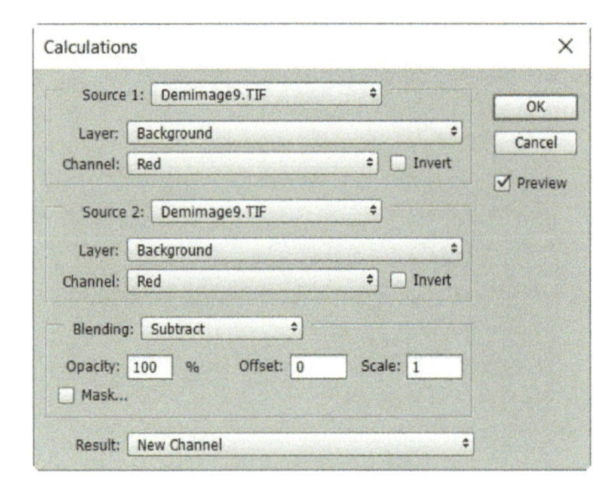

Figure 4.3 Subtract selected as Blending option.

images. These differences are usually small, or in many pixels, there will be no difference at all, resulting overall in a very dark image that is difficult to evaluate. The offset is a valuable tool because it allows one to create a comfortable gray matrix against which the result of the subtraction can more easily be viewed.

- Re-set the *Offset* value at 90. The offset can be any value chosen by the user. The higher the offset value, the lighter gray the matrix will be. The example below illustrates the value of offset.

Channel Subtraction

One channel may be subtracted from another within the same image, in any color mode. There are obviously no registration issues in this process, since they are channels of the same image. Ideal candidates for this process would be one channel that contains clear and strong fingerprint detail and the obstructing background, and another that displays the background but little or no presence of the fingerprint.

- The image is opened in RGB (Figure 4.4).
- Image > Calculations > Subtraction is selected, with the *Offset* value set at 0.
- Red and blue channels are subtracted.
- Next, the *Offset* value is changed to 90 (Figure 4.5).
- The resulting image with an offset of 90 reveals the result of subtracting the red and blue channels.

Since the release of Photoshop CS3 Extended, it has been possible to create an image stack and auto-align the images. This means that if a disruption in image registration has occurred between the captures of the source images,

Figure 4.4 Image opened in RGB.

they can be realigned quickly and easily. The authors believed initially that this option was merely a contingency for situations in which the tripod was accidentally moved or jostled, breaking the registration of the images. It has become apparent that use of the auto-alignment feature may provide a superior subtraction result even when no apparent disruption occurred, and it should be used in every subtraction undertaken.

Caveat

It is essential to the effectiveness of an image subtraction that when possible, the lighting, camera, and subject position remain unchanged in both images. Image registration must be preserved or acquired prior to subtraction. If the angle of ambient lighting were to be different in the second image, artifacts would be created, even if the registration is preserved. Ideally, the only difference between the two images should be the presence of the fingerprint, scale, and identifying markings in the first one. Auto-alignment of images within Photoshop can bring two misaligned images into registration, but it cannot solve the problem created by changes in lighting between the input images.

Filtration Subtraction

The first image subtractions in the 1990s were, as previously stated, conducted with grayscale cameras. A fingerprint developed with ninhydrin on a piece of Canadian currency in this time period was not possible to evaluate because

(a)

(b)

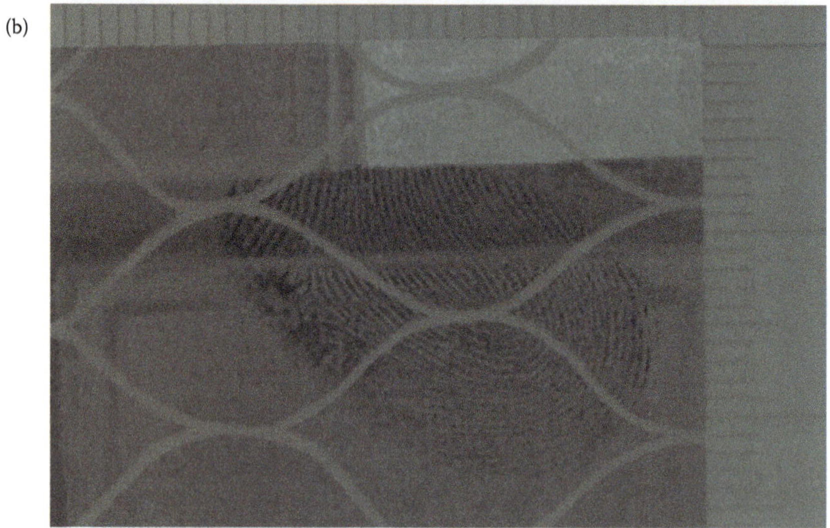

Figure 4.5 (a) Result of subtraction with no offset value. (b) Same subtraction with an offset value of 90.

of significant obstruction due to the engraving pattern on the bill. A grayscale image of the fingerprint and background together was first captured (Figure 4.6a).

Next, a Kodak #33 magenta filter was placed at the camera lens, and a second image was captured. Care was taken to ensure that the position of the camera, lighting, and subject did not change. The ninhydrin fingerprint was eliminated by the magenta filter and did not appear in the second image (Figure 4.6b).

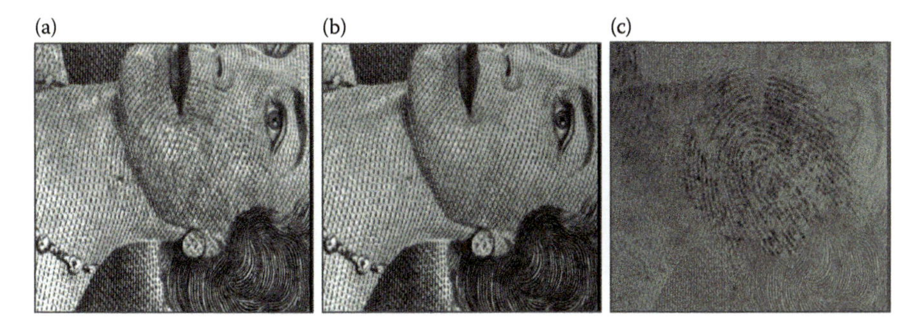

Figure 4.6 (a) Ninhydrin-developed fingerprint on currency. (b) Same image photographed with a Kodak #33 magenta filter to remove fingerprint. (c) Result of subtraction.

The images were subtracted in Photoshop, and the resulting image was evaluated. The background pattern had been suppressed, and the fingerprint detail was clearer and stronger (Figure 4.6c).

Rationale for Erasure of Evidence

The suggestion that a fingerprint or shoe print should be erased in the process of photographing it may seem improper on first reflection, but there are several factors that place this action in perspective.

First, forensic examinations that are completely nondestructive or nonaltering are in the minority. Examination by different wavelengths of light is a notable example, although there have been instances of damage to either the exhibit, fingerprints, or DNA evidence caused by laser or ultraviolet irradiation. Most other fingerprint detection techniques require the exhibit to be immersed in solutions, covered in powder, or exposed to the fumes of cyanoacrylate. These procedures can change the appearance, sometimes drastically, of the exhibit. Inks can run on documents after exposure to organic solvents, rendering the original information unreadable. Obviously, these consequences do not exclude the use of such techniques. The item can be photographed prior to treatment to record its original condition.

DNA extraction, soil analysis, and even the autopsy process *are* invasive examinations that require the destruction or dissection of the sample or exhibit to reveal evidence. The recovery of essential evidence justifies these procedures and, indeed, is not possible without them. There are, of course, conditions to be met when conducting such procedures.

- The documentation, written and photographic, is transparent and complete.
- Every attempt is made through sequential processing to avoid damaging or destroying any other potentially significant types of evidence.

- Damage to and alterations in appearance of exhibits are the result of the desire to recover the best and most complete evidence possible.

There is an option in any human endeavor to complete the easiest or least taxing version of a task, rather than to leave nothing undone, unaddressed or undiscovered. This is referred to metaphorically as the "low-hanging fruit principle," in which only the closest and easiest fruit is picked. Everyone working in the discipline of forensic identification has an ethical obligation to extract the maximum amount of evidence possible from crimes scenes and exhibits. The best evidence dictum can be traced back through British common law to the eighteenth century.

Erasure subtraction is an option for impression evidence that is obstructed and unusable, beyond the reach of less invasive strategies.

Lastly, much impression evidence is destroyed or left behind after examination and photography. Impression evidence within crime scenes is not always amenable to transport. A police agency will diligently protect a crime scene while evidence recovery and recording are underway, only to engage the services of a cleaning company to remove all traces of the crime when they have finished their gathering and recording activities. *What we do is not as important as why we do it.*

Erasure Subtraction

The camera, subject, and lighting are all positioned for an optimal recording of the impression, ensuring that the following criteria are met:

- The scale, subject, and image planes are parallel.
- The subject area of interest is uniformly illuminated.
- The subject is in focus.
- The subject, scale, and markings fill the frame, affording maximum image resolution.
- The aperture that affords best image sharpness is selected.

Today, impression images are routinely captured in RGB color mode, and the filtration strategy no longer has the same value as it did 25 years ago. In any case, the filtration method had limited value when the desired impression was of a complex color (taupe or beige, for example), or when the substrate itself was multicolored. Impressions such as fingerprints and shoe prints are often recorded in media that are very similar in color to what is seen in the substrate. In these instances, the filter would create as many difficulties as it solved, lightning or darkening different areas of the background and creating artifacts in the subtracted image. It is critical to the success of the procedure

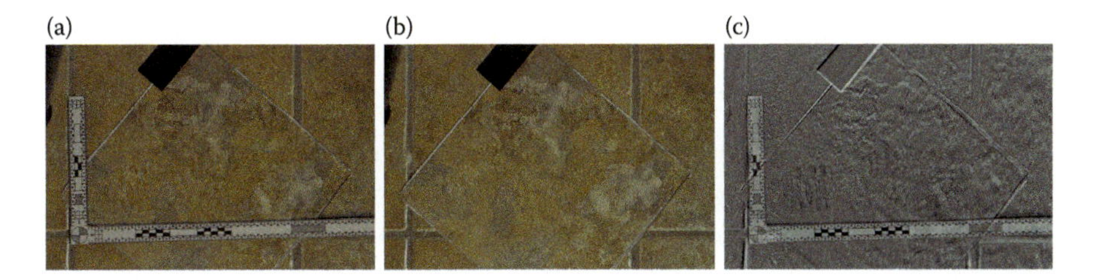

Figure 4.7 (a) Footwear impression photographed with scale. (b) Background photographed after footwear impression erased and scale removed. (c) Subtraction result without auto-alignment.

that the background appear as close as possible to the same in both images. This calls for a more aggressive strategy—erasure subtraction.

The first exposure is taken, recording the impression as clearly as possible against the obstructive substrate. Next, the scale (with impression markings) is removed. The surface is then cleaned thoroughly, in the manner most suitable to the surface and impression medium. For floor surfaces and other nonporous substrates, this is usually a moistened paper towel. In the case of a suspected blood impression, the acquisition of a sample on a swab for DNA analysis can be concurrent with removing the impression. The surface is then wiped with a dry paper towel to confirm that no wet areas remain. These could result in image-destroying specular reflection. The final image of the background is then recorded.

In the example below, the footwear impression is recorded in a medium that closely approximates the colors present in the floor tile design, and it presents a challenge in determining what are bona fide impression minutiae, rather than details in the tile design and color (Figure 4.7a).

In this example, the tripod was moved intentionally after the first exposure and before the second to demonstrate the value of auto-alignment when completing image subtractions. The scale was removed, and the impression was erased with a damp paper towel, followed by a dry one (Figure 4.7b).

Two procedures were followed. The first, procedure A, was conducted entirely in Calculations. This was the only option for subtraction prior to the introduction of the auto-alignment feature (circa 2010). The second, procedure B, utilized the auto-alignment feature.

Procedure A—Traditional Subtraction Method

- Open both images in Photoshop.
- Select Image > Calculations > Subtract.
- Ensure that the two images appear in Source 1 and Source 2 windows, respectively.

- Under Result, select New Document.
- Select OK.

The movement of the tripod between exposures has noticeably caused the images to fall out of register, and the subtraction process has failed to delete the substrate and produce a clear and useful image of the impression (Figure 4.7c).

Procedure B—Subtraction Using the *Auto-Align* Feature in Scripts

- Photoshop is opened.
- Select File > Scripts > Load Files into Stack.
- Select *Browse* to open files for subtraction (Figure 4.8).
- If the files have already been opened, select *Add Open Files.*
- The files will appear in the window.
- The box "Attempt to Automatically Align Source Images" must be checked.
- Select OK.
- Photoshop will align the two source images into a single merged file.
- The checkerboard areas around the edge of this file indicate the degree to which the source images are out of register (Figure 4.9).
- Select Image > Calculations.
- There are four options for how to proceed with the subtraction:
    - Red channels
    - Green channels
    - Blue channels
    - Gray (all channels)
- It is important to explore the subtraction in each of these options. The clearest and strongest result may be found in any of these possible combinations. The examples below are the results of subtracting

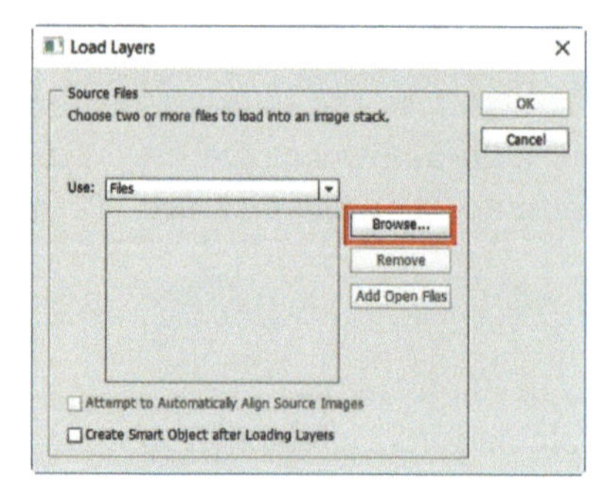

Figure 4.8 Load Layers window showing Browse option.

Figure 4.9 Result of the auto-align feature; the checkered area around image reveals how far the two images were out of register.

the red channels, green channels, and blue channels, respectively, of the two images (Figure 4.10a–c).

- The subtraction of green channels is subtly but distinctly clearer than the red or blue channels. As mentioned in Chapter 3, the green channel contains more information that the other two, and if another channel is equally or almost as clear, the green channel is a wise choice.
- After contrast adjustment in Levels, the final image reveals a substantial increase in signal-to-noise ratio over the original image (Figure 4.11).

Subtraction by Pretreatment Capture

When an exhibit with a very complex or assertive background, or one displaying a variety of different colors, is encountered, significant obstruction of fingerprint detail can be expected, even when the chemical treatment is successful in developing ridge detail. In these situations, pretreatment

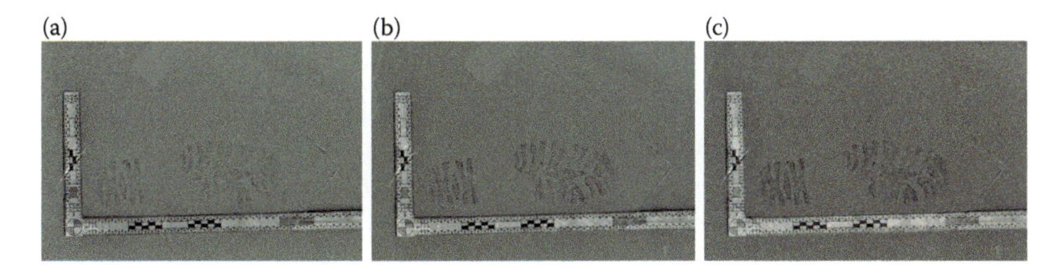

Figure 4.10 (a) Result of auto-aligned subtraction using red channels. (b) Result of auto-aligned subtraction using green channels. (c) Result of auto-aligned subtraction using blue channels.

(a)

(b)

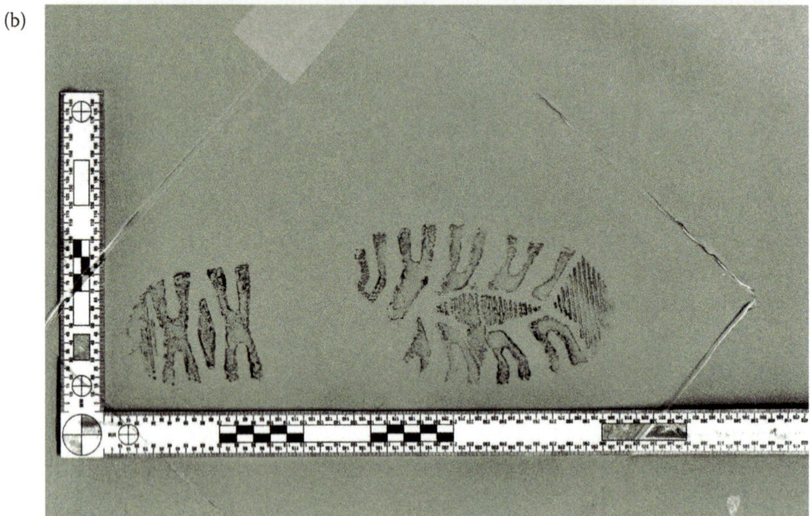

Figure 4.11 (a) Original image. (b) Final subtraction result.

photographic capture of the surface can be beneficial. This procedure is not practical or even possible for all evidence encountered, but the gravity of the investigation and the potential significance of a single piece of evidence may make it worthy of consideration.

The following example is a piece of U.S. paper currency, a $5 bill, selected because of the high level of obstruction posed by the background engraving (Figure 4.12).

- The technique selected for this exhibit (after preliminary light examination) is 1,2 indanedione (ID).

Figure 4.12 $5 bill (U.S.) displaying obstructive background.

- Prior to ID treatment, the area of interest was photographed under the light source that would ultimately be used to examine for results, in this case, a TracER 532 nm laser (Figure 4.13).
- The exhibit was then removed from its place in front of the camera (thereby breaking registration), treated with ID, and placed in a heat press for 10 seconds. The area of interest was then re-photographed under laser illumination. Strong ridge detail can be seen in the image, but it remains partially obscured by the pattern of the engraving on the bill (Figure 4.14).
- The two images were auto-aligned in Photoshop and subtracted in Calculations, following the procedure previously described in this

Figure 4.13 Background photographed with TracER 532 nm laser with a barrier filter, prior to chemical treatment.

Figure 4.14 Indanedione treatment reveals ridge detail, partly obscured by background.

> chapter. Subtraction of the red channels was selected. Completion of auto-alignment demonstrates, in the merged file, the degree to which the two images were out of register (Figure 4.15).

The subtracted image in grayscale reveals an increase in the strength of the image, but there is still obstruction from the background (Figure 4.16).

The subtracted image was opened in Image-Pro Premier. Selected areas displaying similar repeating elements (periodic noise) were edited individually in fast Fourier transform (FFT) (see Chapter 6).

Figure 4.15 Auto-aligned images 4.13 and 4.14.

Figure 4.16 Result of images 4.13 and 4.14 subtraction.

Removal of pattern elements in the background that survived the subtraction process has increased the clarity of ridge detail in several areas of both impressions. A comparison between the fingerprint detail after indanedione and the subtracted image can be made (Figure 4.17a,b).

Frequently in such cases, the improvements may be subtle and require careful analysis to be fully evaluated.

Finger impressions recorded in blood on metallic paint are on the threshold of visibility and are not suitable for analysis (Figure 4.18).

Multiple specular reflections from the surface will grow more obtrusive if attempts at contrast adjustments are made. The impressions were photographed with a Rofin Polilight 415 nm forensic light source, using an orange barrier filter (Figure 4.19a).

The ridge detail is slightly stronger, but the specular reflection issue remains. A second image was recorded after the blood impressions had been removed with distilled water and swab (Figure 4.19b).

The images were auto-aligned and subtracted in Photoshop, using the green channels. The resulting image was then adjusted for contrast. The ridge detail has been significantly strengthened, and no trace of the metallic reflections can be seen (Figure 4.20).

A footwear impression on a tile floor is recorded in an unknown dried gray medium, possibly mud. The impression is cleaned from the surface

Figure 4.17 (a) FFT editing removes periodic noise. (b) Original image after treatment with indanedione.

with a damp towel, the scale removed, and a second image was recorded (Figure 4.21a).

Subtraction without auto-alignment results in an improved image, but the impression details are still obstructed by substrate noise (Figure 4.21b).

When the images are auto-aligned through Scripts, much of this noise is removed, leaving a clearer image of the impression (see bottom image of Figure 4.21b). This is particularly visible in the enlarged details of the respective images (see Figure 4.21c).

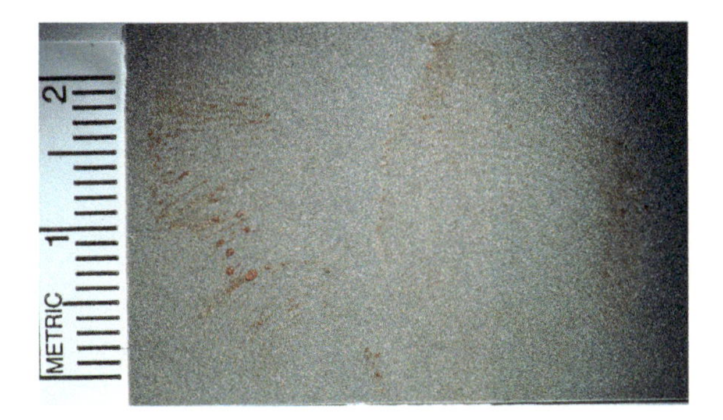

Figure 4.18 Bloody fingerprint impression on the surface of a metallic silver vehicle.

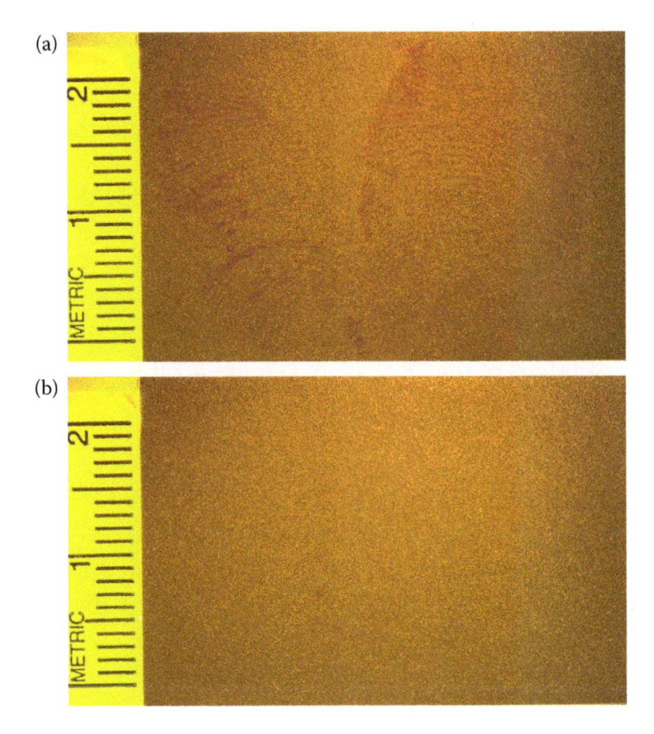

Figure 4.19 (a) Bloody fingerprint impression photographed (using tripod) under 415 nm forensic light source with an orange barrier filter. (b) Substrate photographed (using tripod) after impression removed under a 415 nm forensic light source with an orange barrier filter.

Case Example 1

The subtraction technique is not difficult to perform, but each step must be done carefully and precisely to obtain an optimum result. Partial ninhydrin fingerprints on a check were obstructed by the background pattern in Figure 4.22, and erasure subtraction was attempted. After capturing the initial image

Figure 4.20 Subtraction result.

of the fingerprints, background, and scale, the Ruhemann's purple (ninhydrin stain) was removed with a weak solution of bleach in distilled water. The removal of the stain is accomplished chemically rather than by abrasion. Too much bleach solution was applied, resulting in rippling of the paper, as evidenced by the darker areas.

The overapplication of the bleach solution can create unnecessary and unwanted artifacts in the resulting image (Figure 4.23).

The result of the subtraction displays stronger ridge detail, but closer scrutiny reveals periodic noise from the background printing on the check (Figure 4.24a). FFT editing (see Chapter 5) removes much of this noise, facilitating the analysis of the fingerprint (Figure 4.24b).

Case Example 2

In 1993, a convenience store became the scene of a robbery/murder committed by three subjects. Identification officers processed a plastic laminate countertop bearing a woodgrain pattern with fingerprint powder and developed a footwear impression. Lifting of the impression with clear tape was attempted and placed on a contrasting backing card, but it was readily apparent that much of the impression had not been captured on the tape, and the lift was unusable. The impression was re-dusted with a different powder and lifted with frosted tape. This attempt was equally unsatisfactory.

The countertop now became the subject of erasure subtraction. A digital image of the surface (twice powdered and twice lifted) was captured, with scale and markings (Figure 4.25a).

The scale and markings were removed, and the dusted impression was carefully and thoroughly removed from the countertop, ensuring that the respective positions of camera, subject, and lighting were unchanged. A second image was recorded of the bare countertop surface. The two images were subtracted in Photoshop (Figure 4.25b).

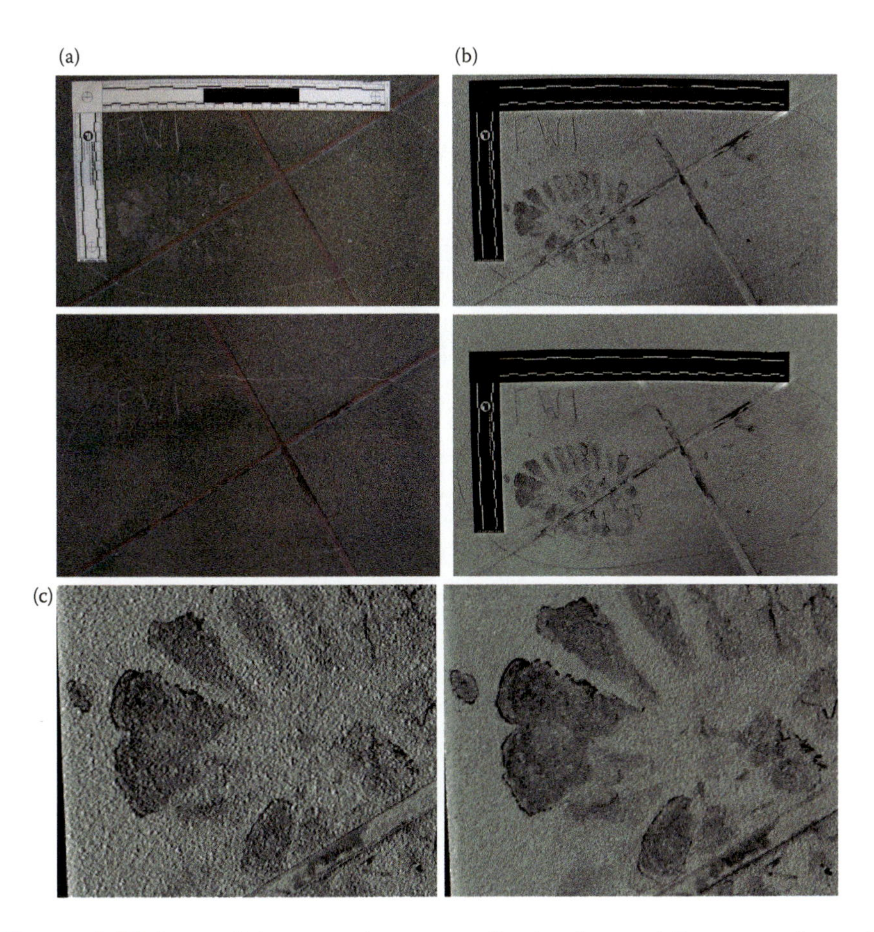

Figure 4.21 (a—top) Footwear image on tile. (a—bottom) Footwear cleaned off surface; note the wetness still present in the grooves of the tile. Ideally, care should be taken to clean substrate and allow it to fully dry. (b—top) Subtraction result without using the auto-align feature in Scripts > Load Files into Stack. (b—bottom) Subtraction result using the auto-align feature in Scripts > Load Files into Stack. (c—left) Subtraction without Scripts. (c—right) Subtraction achieved a clearer image using Scripts to auto-align images before subtraction.

The resulting image displayed a far stronger and more detailed depiction of the shoeprint than either of the lifted versions. Although no footwear was ever recovered during the investigation, the size, make, and model of the shoe responsible for the impression was determined. This information was critical in identification of the shooter in the incident by eliminating the other two participants in the robbery. The subtraction evidence was tendered in court without issue, and a guilty verdict ensued [4].

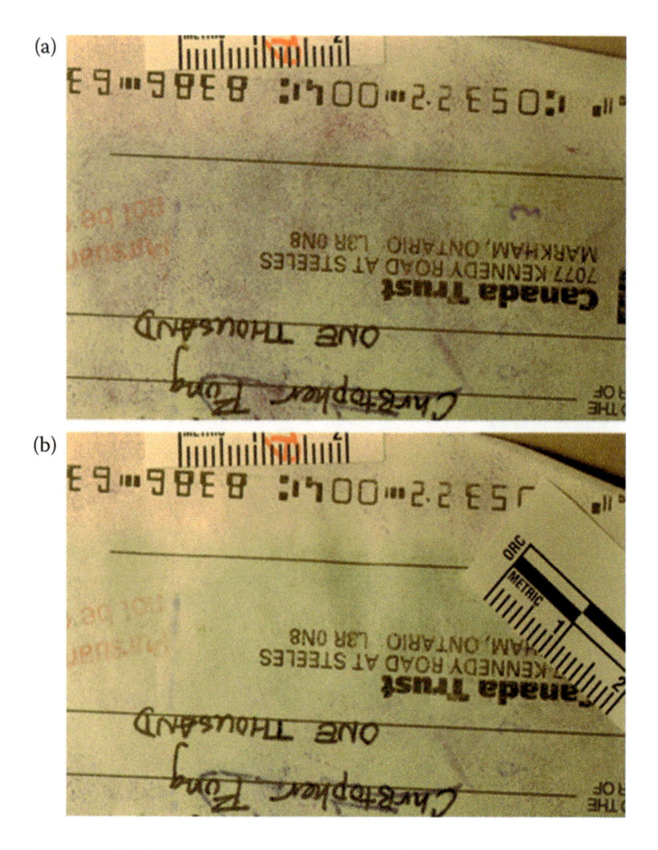

Figure 4.22 (a) Ninhydrin print on a check. (b) Photo of the check after removal of print, showing some damage to the paper.

Figure 4.23 Resulting image.

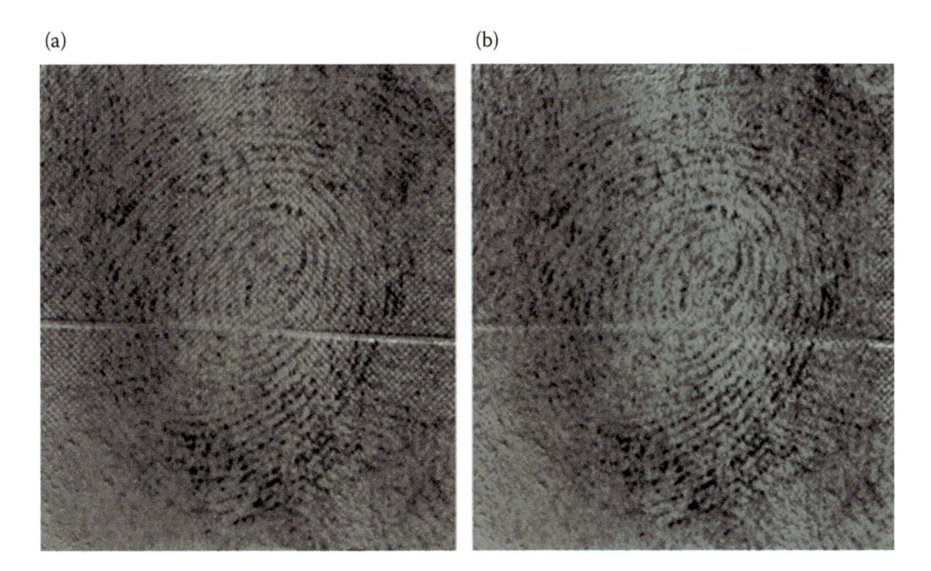

Figure 4.24 (a) Close-up of substrate noise. (b) Noise has been muted using FFT for pattern removal.

Summary

Image subtraction is one method for removing background obstruction in impression images. It is particularly applicable in cases when the desired impression is:

- Borderline in content and clarity.
- Obscured by multicolored substrate details and/or busy patterns.

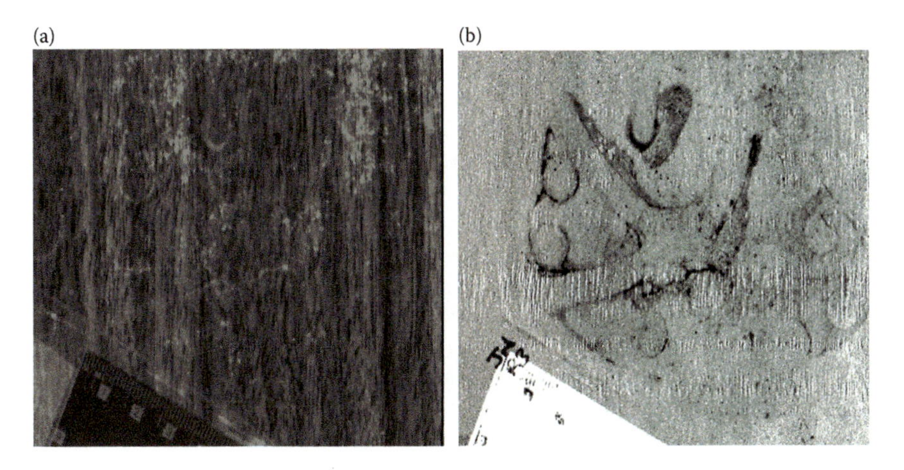

Figure 4.25 (a) Image captured with scale. (b) Results after subtraction. (Courtesy © Queen's Printer for Ontario 2017. Reproduced with permission. All rights reserved.)

Ultimately, there is rarely, if ever, only one method that will accomplish signal-to-noise optimization in any given image.

Focus Stacking

There are two traditional methods of photographing fingerprints or other impression evidence on curved or irregular surfaces. The first is stopping down, using a smaller aperture, which increases depth of field. The second is to increase camera-to-subject distance, making the area of interest smaller in the field of view. Both measures increase the apparent depth of field, and each has limitations, which are described below.

Under scrutiny, the difference in image quality between images recorded with focus stacking at the sweet spot aperture and those at smaller aperture openings may be subtle, and a question arises. Why go to this extent for a minimal increase in sharpness?

The Need for Optimal Image Quality

The goal of any forensic photographer should be to ensure that every evidence capture is the clearest, sharpest, and most complete possible. As is common in forensic photography, such increases may not matter as much when a fingerprint features clear and extensive detail, but they can have great consequence when the impression, or a part of it, is on the threshold of comparison value. This is particularly true when the image prepared for the analysis, comparison, evaluation, and verification (ACE-V) process is subjected to other optimizing processes such as contrast adjustment.

"Garbage in, garbage out" has been a digital mantra for as long as computers have existed. It goes without saying that we strive for top image quality in all forensic photography, but nowhere is it more essential than in impression recording (fingerprints, footwear, etc.), for the following reasons.

An image of this type may be:

- Subjected to extensive optimization procedures in Photoshop or similar software.
- Enlarged significantly, with the attention focused on the clarity of one or two specific minutiae.
- Searched in an AFIS system.
- The pivot point of an entire investigation.
- The subject of intense, adversarial scrutiny and challenge in court.

A fingerprint of exemplary quality containing many minutiae is somewhat more forgiving of small lapses in image quality than one of borderline area and clarity, which requires all the photographer's skill to record it faithfully and completely.

Aperture Selection

The importance of aperture selection is perhaps less well understood than other exposure factors. A series of photographs of identical exposure can be taken, each at different aperture settings, and each will have different properties aside from depth of field. Two factors that should be considered in the choice of an aperture are lens aberrations and diffraction.

Lens Aberrations

A photographic lens receives light from the subject and, in theory, refracts this light to focus on the image plane, resulting in a sharp, focused image. In reality, although modern lenses are corrected for defects or limitations that prevent them from doing this perfectly, the effects of almost all lens aberrations increase in severity toward the periphery of the lens and are least problematic close to the lens axis. The most common of these is chromatic aberration [5].

Additionally, the larger the aperture used, the shallower the depth of field. These two factors would argue against selecting wide open (the largest aperture setting on the lens) for everyday use. There are photographic assignments and strategies that require wide-open aperture photography, but recording of impression evidence is not usually one of them.

Diffraction

As one decreases the aperture size, the effects of lens aberrations diminish, and the depth of field increases, both good things. However, another issue arises. When light strikes the edges of an opening (like the iris in the lens), it is bent or diffracted in ways that reduce the sharpness of the image. As the size of the lens opening decreases, the percentage of diffracted light increases, commensurately reducing sharpness and the resolution of fine detail. Using the smallest aperture setting on the lens would afford the greatest depth of field but at a significant cost in image sharpness and fine detail [6].

A Canadian $20 bill was photographed at apertures F8 and F32 (Figure 4.26).

Enlarged details of the two images clearly reveal the difference in resolution and sharpness (Figure 4.27).

Figure 4.26 (a) Image photographed at F8. (b) Image photographed at F32.

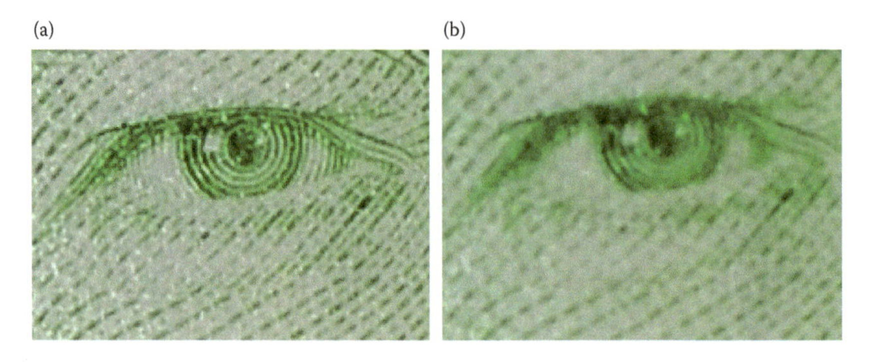

Figure 4.27 (a) Close-up of part of the image photographed at F8. (b) Close-up of the same part of the image photographed at F32.

The Sweet Spot

Clearly, the aperture that provides optimum image sharpness and quality lies between these extremes and is often referred to by photographers as "the sweet spot" [7–9]. It is more of a range than an exact location, but the consensus is that the sweet spot lies two to four stops from wide open. On a Nikon 2.8 60-mm macro lens, for example, the sweet spot is in the range of F5.6 to F8. This certainly does not mean that other settings on the camera shouldn't be used for specific tasks, but using a default aperture for optimum sharpness when photographing fingerprints and other impressions is a sound practice. It must be remembered that sharpness and focus are two different things.

Depth of Field

Light rays come to a point of focus on only one plane, the only one in sharp focus. Objects slightly in front of and behind the plane of focus appear to the human eye as sharp because it cannot discern small differences in precise focus. This range is referred to as depth of field, and it is affected by the following factors:

- The smaller the aperture, the greater the depth of field.
- The shorter the focal length of the lens, the greater the depth of field.
- The greater the camera-to-subject distance, the greater the depth of field.

Depth of field is extremely shallow in macro photography. When recording small subjects such as fingerprints at close range, the human eye is much more sensitive to lack of focus, and the depth of field is measured in fractions of millimeters [10]. Using a Nikon 60-mm 2.8 macro lens with an aperture setting of F8 (author's designated sweet spot) minimizes the effects of both aberrations and diffraction to provide high image quality and works well on flat subjects. However, fingerprints are occasionally developed on curved or uneven subjects and require a greater depth of field than required for a flat plane.

An aperture setting of F8 in macro photography commonly fails to provide sufficient depth of field for a subject on a curved or uneven surface. When the aperture is changed to F22, F32, and F40, there is a progressive increase in depth of field but also a loss of sharpness, particularly in the areas farthest from the center of the image.

In focus stacking [11,12], multiple images are captured of the subject, each focused on planes between the nearest and farthest planes of the subject, each at the lens' sweet spot aperture for optimum image quality (Figure 4.28).

These images are then opened in Photoshop, where the area of sharp focus in each can be combined into one seamless image. All areas of the subject will be in sharp focus at the aperture providing maximum image quality.

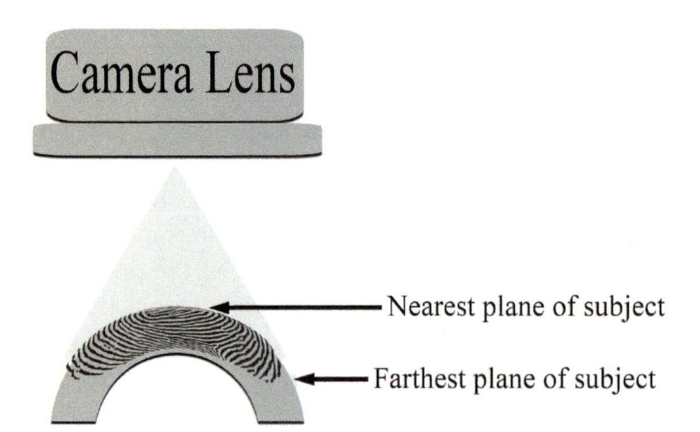

Figure 4.28 This diagram illustrates the different planes of a subject on a curved surface that present a focusing challenge in macro photography.

The example that follows is a seashell with an extreme depth of field requirement, greater than any likely situation encountered in a fingerprint or shoeprint. It has been photographed at F8, with only the closest edge of the shell in sharp focus and the remaining portion of the subject blurred. Next, 20 exposures were taken, again at F8, each focused on one of the planes between the nearest and farthest parts of the subject (Figure 4.29a,b).

A fingerprint is recorded on a cylindrical vial with a short radius of curvature, creating a significant depth of field challenge. When photographed at F8 (chosen sweet spot aperture), the lack of focus in the outer edges of the subject is apparent (Figure 4.30a–c).

Changing the aperture to F22 is an improvement, bringing the edge detail of the fingerprint into sharper focus but at a cost. Diffraction at the smaller aperture has reduced the sharpness and resolution of the fine detail. The focus stacking result, however, displays both sharp focus and resolution of fine detail throughout the subject, as can been seen in the comparison of the enlargements at F22 and the focus stacking result (Figure 4.31).

Focus Stacking Procedure

Focus stacking is also known as focus blending or focal plane merging. The camera, subject, and lighting are positioned to best record the impression, bearing in mind the measures listed earlier in this chapter. The aperture is set at F8, and the focus is in manual mode.

- The focus is set on a point slightly behind the furthest plane of the recorded impression. This is the first exposure and ensures that the portion of the impression most distant from the camera is captured in sharp focus.

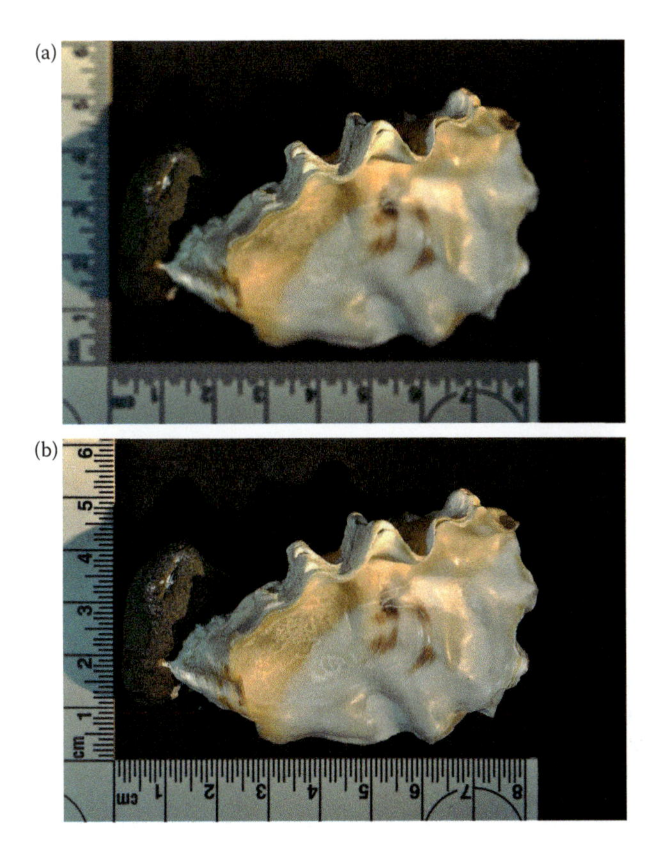

Figure 4.29 (a) A seashell photographed at F8, the ideal aperture for this camera and lens; only the closest part of the shell is in sharp focus. (b) The result of the focus stacking procedure, having taken 20 images all at F8, changing the plane of focus in small increments between each shot.

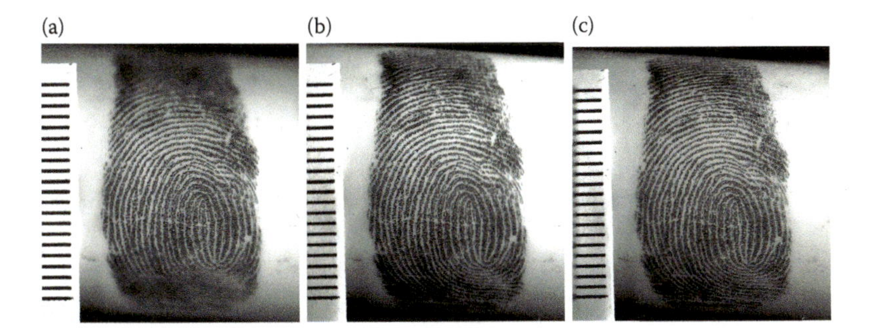

Figure 4.30 (a) Photographed at sweet spot F8. (b) Photographed at F22—diffraction aberration is visibly present. (c) Focus stack results show the image in sharp focus, with no diffraction aberration present.

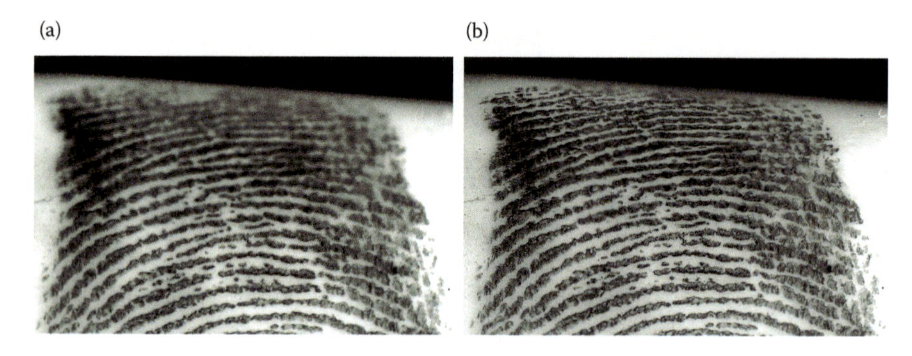

Figure 4.31 (a) Close-up of image shot at F22. (b) Close-up of focus stacking result.

- The focus ring is then adjusted slightly to bring the plane of focus slightly closer to the camera, and the second exposure is recorded.
- This process is repeated (perhaps 15–20 exposures) until the plane of focus is just slightly in front of the nearest portion of the impression to the camera. This results in the entire impression being out of focus, ensuring that all planes of the impression have been recorded in focus. The greater the depth of field requirement, the more exposures need be captured.
- The camera card is then connected to Photoshop.
- Select File > Scripts > Load Files into Stack.
- In the open window, select *Browse,* and highlight the source images for the stacking process.
- Check the box *Attempt to Automatically Align Source Images,* and press OK. It may take several minutes for Photoshop to complete this task, depending on the number of exposures.
- When the alignment process is complete, select the Layers tab to the right of the working window. Using the Shift key and cursor, highlight the layers of all source images.
- At the top of the Photoshop display, select Edit > Auto-Blend Layers > Stack Images. Photoshop then selects the areas of sharp focus from each of the source images and blends them into one seamless image. Again, this process may take several minutes, depending on the number of source images and the abilities of the computer.

The resulting image displays all areas of the impression recorded in sharp focus, each captured at the f-stop that provides maximum image quality.

Software programs can offer very precise focus stacking, as well as many other features, and operate with the camera tethered to a laptop computer [13,14]. There are many articles and tutorials on the Internet dealing with this technique.

High Dynamic Range (HDR) Pro

Dynamic range refers to the luminance values between the brightest and the darkest part of the subject. HDR Pro is a feature in Photoshop that combines multiple images recorded at different exposures, selects the areas in each that are most correctly exposed, and blends them seamlessly into a single image.

In most cases, forensic photographers encounter subjects with a compressed dynamic range, subjects of low contrast, in which the eye has difficulty discerning details. The resulting images are routinely adjusted in Photoshop to expand the dynamic range. Occasionally, however, the range of luminance values is too high for the camera to capture in one image. If the darkest part of the image is properly exposed, the highlight areas are washed out. Exposures that capture detail in the brightest part of the subject may result in the shadow detail being lost. The fingerprint below on a DVD disk has been intentionally unevenly illuminated to illustrate the HDR Pro process. The initial exposure to record the ridge detail in the shadows results in gross overexposure and washout of the detail in the highlights (Figure 4.32a).

Conversely, exposing for the highlight detail leaves the remainder of the fingerprint unreadable (Figure 4.32b).

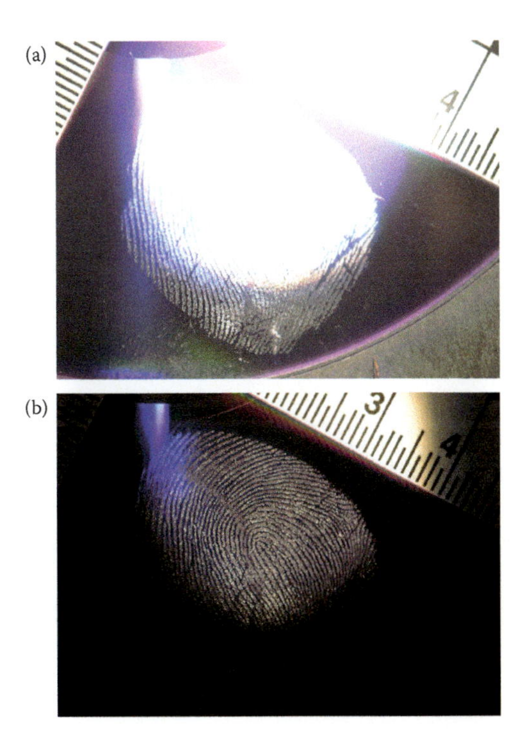

Figure 4.32 (a) Impression exposed to record detail in the shadows. (b) Impression exposed to record detail in the highlights.

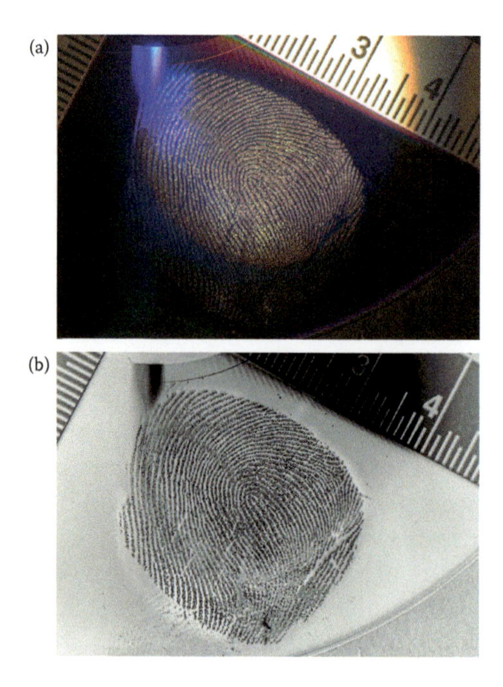

Figure 4.33 (a) Several images of varied exposures are merged to HDR Pro. (b) Image converted to grayscale, inverted, and adjusted for contrast.

A series of different exposures (maintaining the same aperture) is captured and blended in HDR Pro, resulting in a compressed dynamic range in which all the ridge detail is visible (Figure 4.33a).

The image is converted to grayscale, inverted, and adjusted for contrast using the Lasso tool and Levels (Figure 4.33b).

HDR Pro Procedure

If HDR Pro is to be followed, it is strongly recommended that a tripod or copy stand is used.

- One image of the subject is recorded at the metered exposure.
- Two exposures are taken at two stops under and four stops under, respectively.
- Two further exposures are taken at two stops over and four stops over, respectively. The exposure adjustment should be made in shutter speed, using the same aperture setting for all exposures.

The five images are opened in Photoshop.

- Go to File > Automate > Merge to HDR Pro.
- Add open files to the window and select OK.

- A new window will open, *Merge to HDR Pro,* displaying the source images and the HDR blended image, with adjustment controls on the right side of the window. When satisfied with the image, select OK.

The blended image will appear, displaying the compressed dynamic range, composed of the best areas of exposure in each of the source images.

The Power of RAW

Alternatively, if the image is in RAW format, the channels are 14–16 bits rather than 8 bits as recorded in tagged image file format (TIFF) or JPEG, resulting in a vastly increased range of luminance values (see Chapter 2, pages 45–53). These can often be adjusted in RAW, with a similar or even superior result to that obtained in HDR Pro.

Techniques such as 1,8-diazafluoren-9-one (DFO), indanedione, and Rhodamine 6G are such intense fluorescers that the fingerprints they reveal may present high dynamic range images that cannot be adequately recorded, or at least displayed, in a single image. This is a powerful argument for recording impression images in RAW.

A subject similar to Figure 4.33 with even greater exposure disparity was photographed in RAW and JPEG mode in camera (Figure 4.34a).

The JPEG image was adjusted in Levels. Further attempts to reveal shadow detail in Levels resulted in degradation of the highlight information (Figure 4.34b).

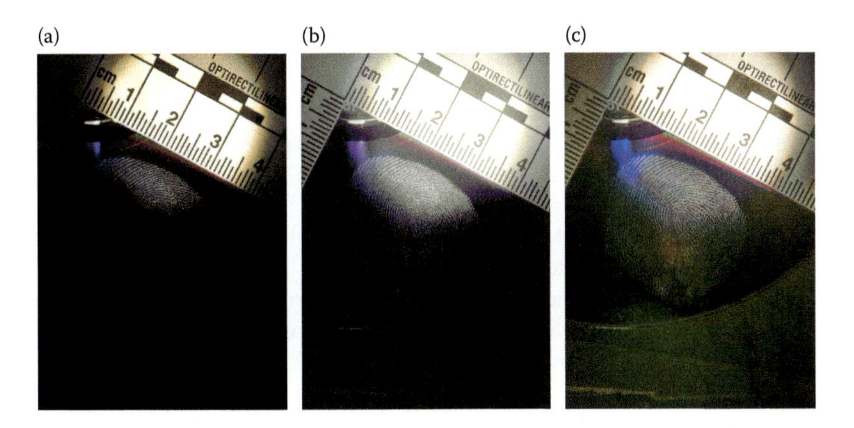

Figure 4.34 (a) Shadows of this image are very underexposed, and detail cannot be visualized there. (b) JPEG version captured, adjusted with Levels. (c) RAW version captured, adjusted in Camera Raw for Highlights, Shadows, and Exposure.

The RAW image was adjusted in Highlights, Shadows and Exposure, resulting in the visualization of the ridge detail previously hidden in the shadow area (Figure 4.34c).

Summary

Most image optimizations are successfully completed on a single image. On occasion, however, the proactive acquisition of two or more images can add greatly to the options for optimizing signal-to-noise ratio. Techniques such as image subtraction and focus stacking frequently result in a final image that is clearer and sharper than one obtained from a single source image.

Review Questions

1. How can one use color modes to complete a channel subtraction?
2. What is meant by optimizing signal-to-noise ratio?
3. How does diffraction reduce the sharpness of an image?
4. What is meant by the "sweet spot" in aperture choice?
5. When conducting an erasure subtraction, ideally what elements must be present in the first image and absent in the second?

References

1. B. Dalrymple and T. Menzies, Computer enhancement of evidence through background noise suppression. *Journal of Forensic Sciences*, 39(2), pp. 537–546, 1994.
2. B. Dalrymple, Cost effective forensic image enhancement wins acceptance, Videometrics VI, *Proceedings of SPIE/IST Conference*, San Jose, CA, 1999. (Reprinted in *Advanced Imaging Magazine*, August 1999.)
3. B. E. Dalrymple, Optimized digital recording of crime scene impressions. *Journal of Forensic Identification*, 52(6), pp. 750–761, 2002.
4. J. Norman, OPP Forensic Identification Services (ret.), personal communication, 2016.
5. K. Rockwell, Glossary of lens terminology, available at http://www.kenrockwell.com/tech/lenstech.htm#ca
6. What is lens diffraction? *Photography Life*, March 2017, available at https://photographylife.com/what-is-diffraction-in-photography
7. Rules of thumb—Finding your lens' sweet spot, B&H Photo video, available at https://www.bhphotovideo.com/explora/content/rules-thumb-finding-your-lens-sweet-spot
8. How to identify your lens' sweet spot, Digital Photography School, available at https://digital-photography-school.com/find-your-lens-sweet-spot/

9. Light stalking, the sweet spot lens setting that will give you sharper photographs, available at https://www.lightstalking.com/sweet-spot/
10. Depth of field calculator, PhotoPills, available at www.photopills.com/calculators/dof
11. B. Dalrymple and J. Smith, Focus stacking in Photoshop—Depth of field optimization in macrophotography. *Journal of Forensic Identification*, 64(1), pp. 71–83, 2014.
12. Digital Photography School, A beginner's guide to focus stacking, available at https://digital-photography-school.com/a-beginners-guide-to-focus-stacking/
13. Helicon Soft, available at http://www.heliconsoft.com/heliconsoft-products/helicon-focus/
14. Control my Nikon 5.3, available at https://www.controlmynikon.com/

5 Fast Fourier Transform Background Pattern Removal

Science is a way of thinking much more than it is a body of knowledge.

Carl Sagan

This chapter describes the process known as fast Fourier transform (FFT). Repetitive background patterns such as dots or lines often obscure detail in fingerprints or other impressions. They can be inextricably blended with the ridge detail, making analysis of the fingerprint difficult or impossible. These obstructions are often difficult or impossible to remove in the spatial domain with conventional strategies. They may be composed partly or completely of grayscale values, reducing the effectiveness of techniques based on color. In FFT, the image is converted to the periodic or frequency domain, in which the image data are organized and displayed as a function of frequencies rather than position in space.

To make the complex mathematics understandable, the FFT process has been compared to "unbaking" a cake [1]. Of course, one cannot remove one specific ingredient of a baked cake. Following this analogy, however, the components of the cake (chocolate, flour, sugar, etc.) are separate unique frequencies of different strengths. Fourier transform would display the components in the frequency domain, where they would occupy different positions. In this domain, one ingredient can be removed from the whole because it can be isolated. The sugar can be removed from the cake.

Repetitive pattern generates unique energy spikes or signatures that appear in different locations in the frequency domain. These spikes can be removed without altering the fingerprint data. The edited frequency display is then converted back to the spatial domain, with the interfering pattern diminished or eliminated.

The first forensic application of FFT known to the authors was conducted in 1971 [2]. A murder in San Diego yielded one critical piece of evidence—a bedsheet bearing a blood transfer impression of ridge detail. The weave pattern of the cloth obscured the faint fingerprint, rendering it unsuitable for comparison in its original appearance (Figure 5.1).

The impression was forwarded to the Jet Propulsion Laboratory in Pasadena, California, to areas called the Space Technology Applications Office and the Image Processing Laboratory. A new process for the removal

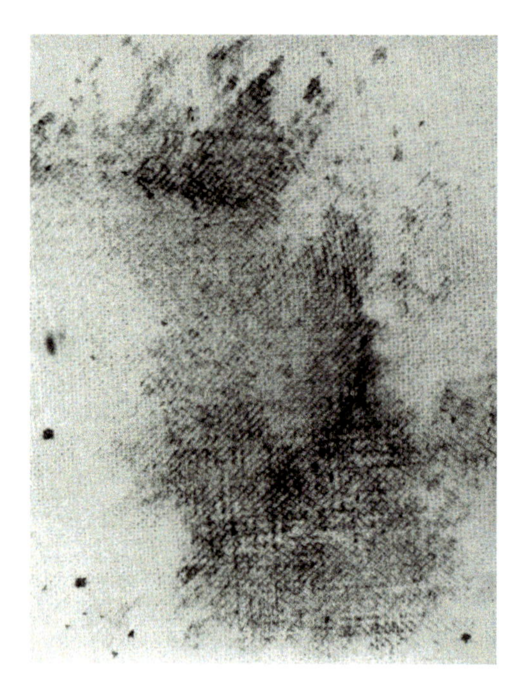

Figure 5.1 Blood transfer of ridge detail on bedsheet. (Courtesy of Charles C Thomas, Publisher, *Scott's Fingerprint Mechanics.*)

of interference in images, specifically a repeating pattern, was in the development stage. Their focus was the clarification of photographs taken by satellites in the space exploration program.

The Fourier transform was first used to create a frequency or periodic display of the image, in which the frequency components of the weave pattern of the fabric were displayed, identified, and removed. It was then applied to transform the image data back to the spatial domain, in which the obstructive weave pattern was gone, and the ridge detail could be seen much more clearly (Figure 5.2).

A comparison of this clarified impression was made with the right palm of the suspect, and the resulting chart illustrated 17 points of agreement (Figure 5.3).

The court ruled the enhanced impression inadmissible because "*the People have failed to establish by a predominance of the evidence, any of the following:*

1. That the scientific principal used by the Jet Propulsion Laboratory in the development of People's 46 (the computer-enhanced photograph) has been generally accepted among experts in the particular area involved.
2. That the technique has been accepted in the particular field of use to which it is applied in this case.

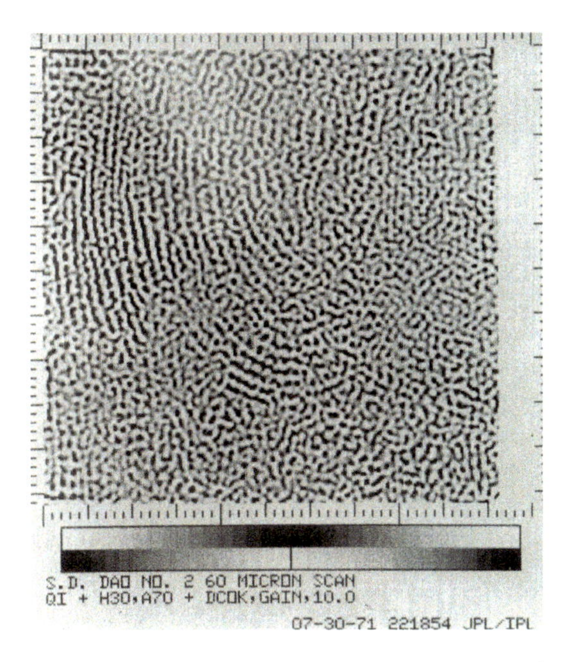

Figure 5.2 FFT transform of image completed by Jet Propulsion Laboratory. (Courtesy of Charles C Thomas, Publisher, *Scott's Fingerprint Mechanics*.)

3. That there is scientific certainty about the results produced by Jet Propulsion Laboratory in this case."

"No latent fingerprint expert was permitted to testify to the value of the enhanced print of the Jet Propulsion Laboratory."

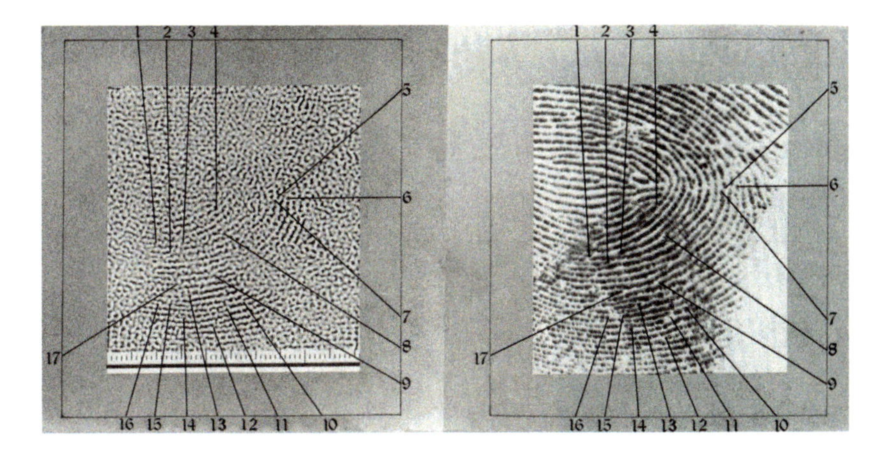

Figure 5.3 Comparison chart with known impression. (Courtesy of Charles C Thomas, Publisher, *Scott's Fingerprint Mechanics*.)

Fast Fourier Transform (FFT) Evolution

Obviously, many things have changed in the intervening 46 years.

- Computers in 1971 were principally confined to government, military, universities, and other large-entity laboratories. The first personal computer (PC) would not be introduced until 1975.
- Computing power and speed were much lower.
- Spatial resolution of images was very low.
- The first operating system for graphics (Windows 1.) was not released until 1985. Complex computer graphics algorithms (such as fast Fourier transform) were conducted in DOS (disc operating system), unintuitive and restricted to those with advanced computer skills and knowledge.
- In 1971, almost no one in the forensic science community (and very few in any other scientific community) were even aware of, let alone comfortable in the application of FFT.
- The FFT enhancement process described in this murder case would have required many hours.

Moving forward to the 1990s, FFT was used to great advantage in both the United States [3,4] and Canada [5]. Despite advances and improvements in computer hardware and software, FFT was still a ponderous, complex, little-known, and intimidating procedure. These were the primary motivators for Erik Berg to create a simpler, more intuitive program utilizing the Fourier transform to remove obstructive pattern in images [6].

Perhaps the most underutilized image optimization technique today is removal of obstructive patterns with fast Fourier transform (FFT), which can routinely make the difference between an indecipherable area of obstructed ridge detail and an identifiable fingerprint. There may be several reasons for this. First, no form of FFT has ever been included in any version of Photoshop, the program most familiar to forensic practitioners, and consequently, it is not automatically within their acquaintance, much less their acquired knowledge sphere or skill base. Several options for FFT as Photoshop plug-ins have been available for about two decades.

Second, the acquisition and use of FFT, either as a separate program or plug-in for Photoshop, represent additional expense, knowledge, and time, the rationale for which may not be immediately obvious and pressing due to reason one.

Third, and arguably the most significant, FFT is somewhat more complex in its function and is less instantly intuitive than other processes such as contrast adjustment, channel blending, and image subtraction. The user will require at least a basic understanding of how a different display domain enables one to perform significant operations on an image that cannot be done in the spatial domain, the normal view of an image.

Software Choices

Today, there are several software options for FFT from which to choose. Each of these offers a slightly different approach in terms of applying the algorithm. Some are Photoshop plug-ins, while others are stand-alone programs. Foray Technologies features a pattern gallery in which the user can identify unwanted patterns detected in the image and drag them into the Trash Can while reviewing the effect of the pattern removal in the preview window. Ocean Systems ClearID consists of a series of sliders, allowing the user to view in real time, the effects of a three-stage adjustment process.

The program preferred by the authors and used to create the images in this chapter is Image-Pro Premier, available at Media Cybernetics. This stand-alone software features a dark periodic display against which the noises spikes are clearly visible and easily edited. It also allows the user to edit selected sectors of the image individually. This option is particularly valuable when the interfering pattern is complex and changing from one part of the image to another.

Those wishing to acquire this capability are encouraged to explore and test the different versions of FFT, to find the software they are most comfortable with.

Anatomy of a Digital Image

It is important to remember that when an image is opened in Photoshop or other processing program, it is a collection of numbers (zeros and ones) that resides on the hard drive or a storage device. The image is normally opened in the spatial display, the configuration at the time of its initial capture, and the subject is seen, be it fingerprint, crime scene, or portrait, as it appeared in life.

Changing the Image Display

In the digital world, we are not limited to only one way of displaying the image data. If one considers the daily use of email as an example, the inbox messages are arranged in chronological order, perhaps with the most recent message on the bottom and the rest above, going back in date and time. It is not necessary or advisable to scan the entire inbox to locate a desired message from a specific person. The display of messages can be quickly changed

Figure 5.4 Different displays of same image content.

to alphabetical order with a click of the mouse to view all messages from a specific person. When the required information has been recovered, another click of the mouse returns the inbox display to the one most suited for daily use. *The inbox data hasn't been changed in content during either conversion. It has been displayed in a different arrangement.* The important thing to remember is that completing certain tasks may require the temporary transformation of the data to a different display.

Similarly, by converting the image to the periodic display, the data are organized as a function of their frequencies, not their original position in space. The result looks nothing like an image. A black rectangle with a bright white cross in the center dividing the display into four quadrants is usual, but as was the case with the change in email display, *the data have not been altered.* All the image data are there, unchanged. This can be verified by simply converting it back to the spatial display.

In another example, examine the different options for displaying shape and color elements. There are circles, squares, and triangles, colored red, green, and blue. They are randomly positioned in the first image. In the next image, they are grouped according to shape, disregarding color. Finally, in the last image, they are grouped according to color, disregarding shape. This is an example of how image data may be displayed in different ways without altering the content (Figure 5.4).

Pattern Signatures

Footwear and fingerprint impressions located on exhibits and at crime scenes are routinely developed with chemistry or powder, or they are simply photographed in situ as they are found. In either case, repetitive patterns such as rows of lines or dots can seriously obstruct the desirable detail. Such patterns are referred to as periodic noise, and they can be difficult to remove with other processing tools.

If there are repeating pattern elements within the image (such as dots or parallel lines), they generate a *signature* in the periodic display, consisting of white dots or spikes (see information that follows). The nature and location of the signature in the periodic display are determined by the uniformity of the pattern—the size, frequency, direction, and gray value of the repeating elements in the image. The key to successful application of FFT is understanding and diagnosing the periodic display of images, and herein lies one of the fundamental guidelines. Once identified, the spikes can be removed, and when the edited periodic display is converted back to spatial (normal) display, the obstructive pattern has either been de-emphasized or removed entirely. As stated earlier in the book, the goal is to optimize the signal-to-noise ratio.

Successful FFT processing can be summed up in this phrase—*remove the spike signatures of the obstructive pattern while leaving the fingerprint signature data untouched.*

There are several useful diagnostic guidelines for accurately identifying the signature of noise in images:

- The more uniform the repeating pattern elements in the image are in size, direction, frequency, distance apart, and gray value, the more focused and easily identified their corresponding signature spikes in the periodic display will be.
- If the repeating elements are on a horizontal axis in the image, the spikes will appear on a vertical (opposite) axis in the periodic display (at 90 degrees).
- The smaller and closer together the repeating elements are, the farther they will be from the dynamic center in the periodic display.

Starting with simple repetitive patterns, examine the images that follow and the corresponding periodic display of each.

We can see the bright, focused signature spikes of the diagonal lines at 90 degrees to their direction in the original image (Figure 5.5).

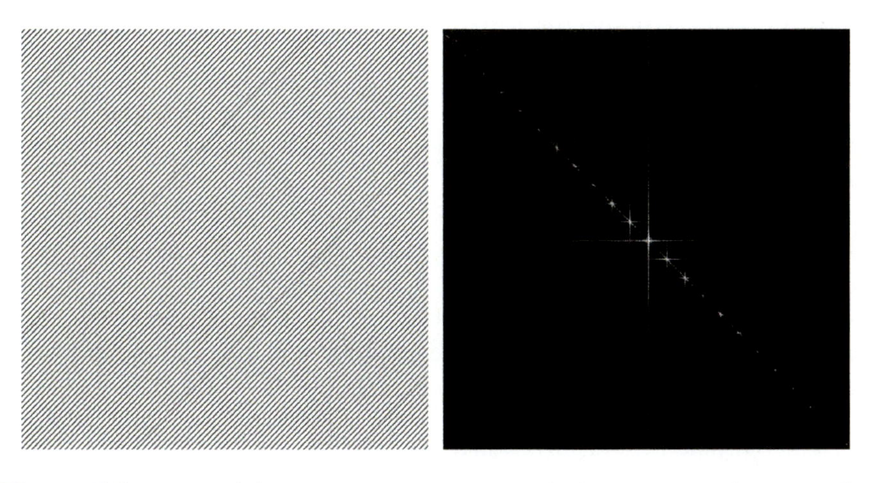

Figure 5.5 Diagonal line pattern in image and the corresponding periodic display.

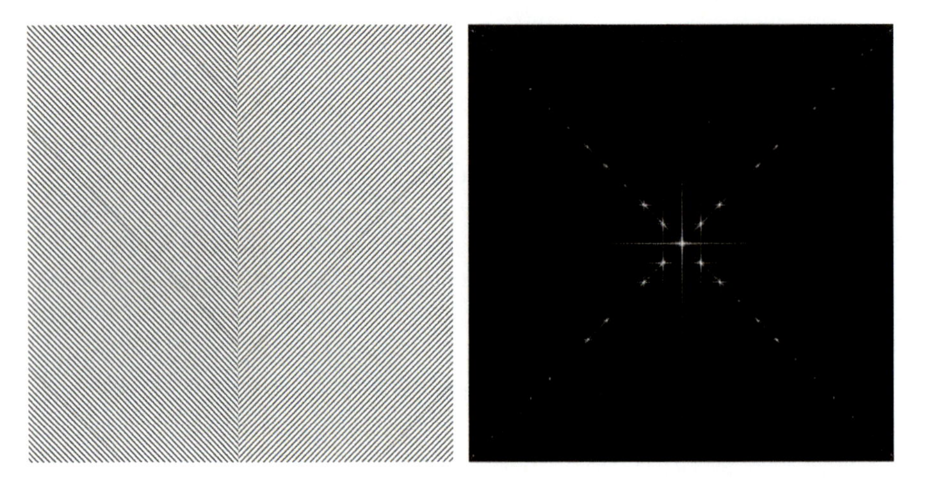

Figure 5.6 Herringbone pattern in image and the corresponding periodic display.

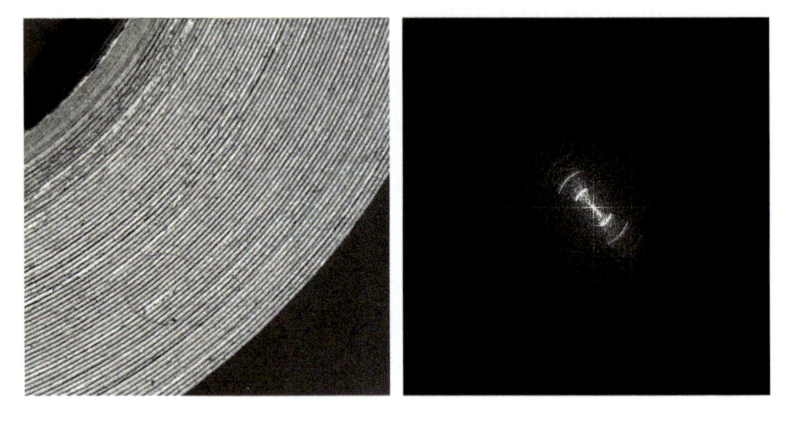

Figure 5.7 Curved parallel line pattern in image (side view of a roll of duct tape) and the corresponding periodic display.

The herringbone pattern features lines on both diagonals, and accordingly, the signature spikes appear along both diagonals in the periodic display (Figure 5.6).

When the parallel lines are curved, the signature spikes are no longer on a single axis. They indicate and represent the range of angles followed by the curved lines in the image (Figure 5.7).

Images that are random and arbitrary in tonal intensity (no repeating pattern) result in a periodic display devoid of signature spikes, as in the natural scene of random rocks and weeds (Figure 5.8).

A fingerprint is distinctly different from a uniform, mechanical pattern of parallel straight lines, as depicted previously, but it is a repeating pattern of roughly parallel lines (ridges) that change direction. The ridges are approximately the same width and distance apart and may vary significantly in gray

Figure 5.8 Image without a repeating pattern; contains no signature in the periodic display.

value. They exhibit a signature in the periodic display but quite unlike that of uniform, mechanical dots or lines. The signature of a fingerprint appears as a haze or smudge rather than a series of single point, focused spikes. A faint fingerprint generates a signature that is very difficult to see at all.

Sharply defined and focused spikes, such as the ones associated with uniform, machine-generated patterns we saw previously, are a dead giveaway for noise in an image. If one acquires a knowledge of the periodic domain and learns the language, so to speak, there is little danger of misdiagnosis. These spikes are routinely found in a different part of the periodic display from that occupied by the fingerprint data, and consequently, they can be edited while leaving the fingerprint unaltered.

Examine the fingerprint that follows and its corresponding display in the periodic domain (Figure 5.9).

Now, examine the image below, which includes both the fingerprint and an obstructive, repetitive pattern. When converted to the periodic display,

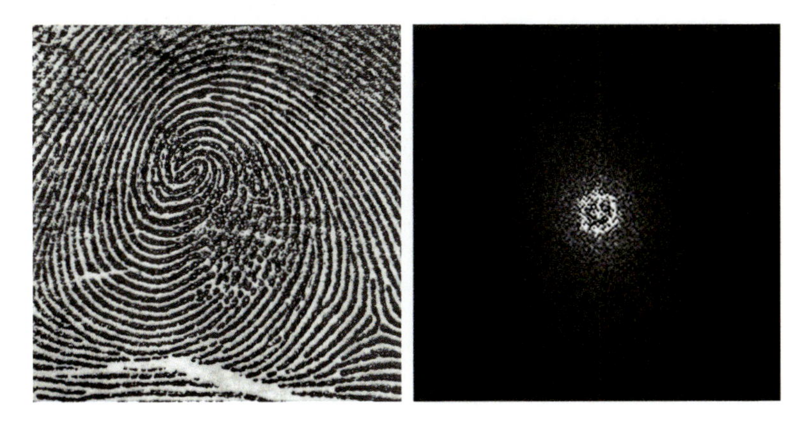

Figure 5.9 Fingerprint and periodic signature.

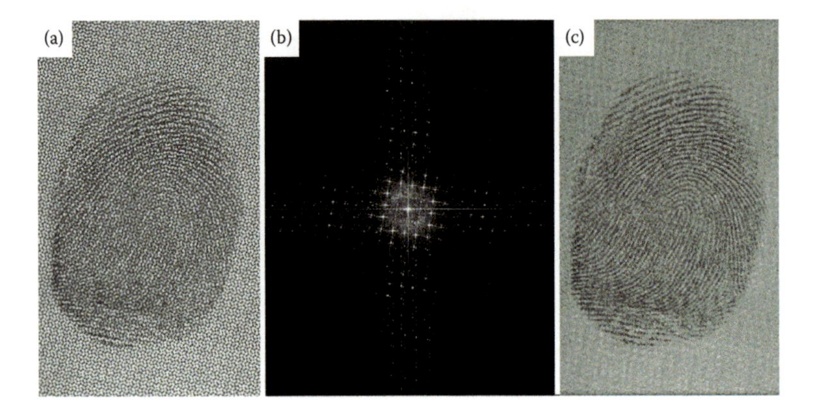

Figure 5.10 (a) Obstructed fingerprint. (b) Pattern spikes (signature) in periodic display. (c) Result of editing.

there are two strikingly different signatures—bright, focused dots representing the mechanical pattern, and the soft, hazy detail of the fingerprint ridges in the form of an elliptical donut. All elements of the pattern signature in the periodic display are removed. When the image is transformed back to the spatial display, the pattern has been suppressed, and the ridge detail can be followed and evaluated much more easily (Figure 5.10).

In the image that follows, a fingerprint has been deposited on Kevlar cloth. The weave of the Kevlar is so obstructive that the ridge detail cannot be evaluated. When the image is converted to periodic display in FFT, however, a signature array of spikes, attributable to the weave pattern, can be clearly seen, as well as the soft, hazy signature of the fingerprint (Figure 5.11).

The "Spike Cut" feature is used to remove these obstructive spikes (see Figure 5.12). When the image is converted back to spatial display, the ridge detail can be seen and evaluated.

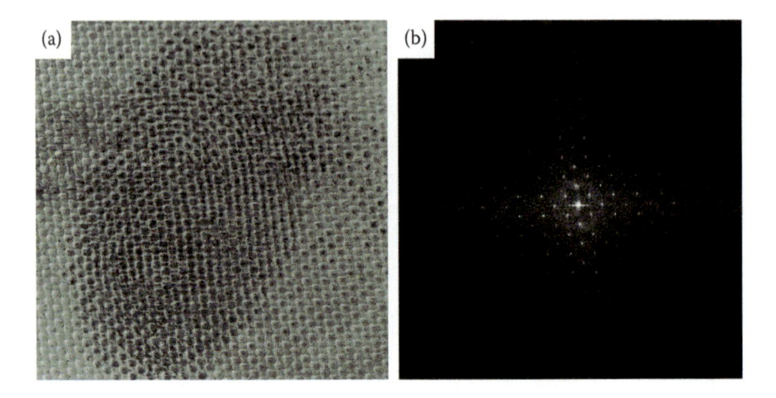

Figure 5.11 (a) Fingerprint on Kevlar. (b) The periodic display showing signatures for both the Kevlar weave and the fingerprint. (Courtesy of Scott Booker, York Regional Police.)

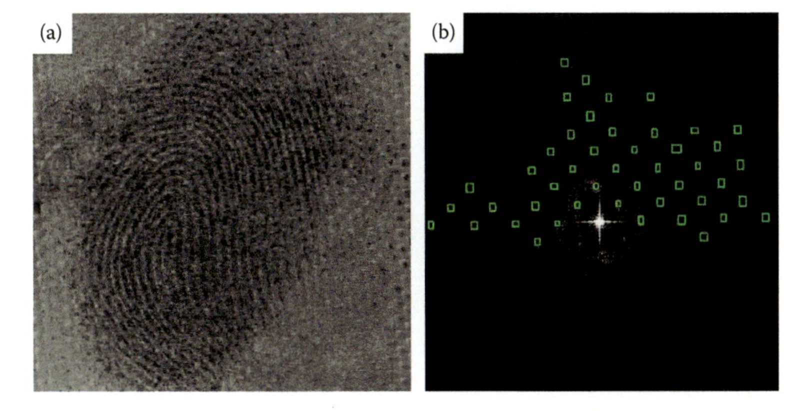

Figure 5.12 (a) Results of editing. (b) Noise spikes outlined and removed.

It is necessary to stress that the fingerprint data have not been changed or altered. They occupy different parts of the periodic display and are not affected when the spikes associated with the Kevlar pattern are edited. The obstructive pattern has been removed, allowing the eye to see the ridge detail that was present but obscured in the original image (Figure 5.12).

Sequential Processing of Images

Both previous examples featured 8-bit grayscale images with one predominant issue—a uniform obstructive background that did not vary significantly within the image area. For this reason, it was possible to diagnose and remove the pattern from the entire image in one operation.

Some images, however, require a sequence of steps that may be completed in different software programs.

Example 1—Channel Subtraction Followed by FFT

For example, the image that follows features a finger impression developed with ninhydrin on Canadian paper currency (Figure 5.13).

The ink of the background printing is a different color than the ridge detail. Bearing in mind that there is seldom, if ever, only one method of accomplishing a successful enhancement, here is one sequence of operations that results in a significant improvement in the signal-to-noise ratio, revealing ridge detail that was initially obscured by substrate noise.

1. First, the image is opened in Photoshop.
   a. The channels are examined for blending or subtraction opportunities.

Figure 5.13 Ninhydrin fingerprint on Canadian currency.

b. When the image is opened in RGB color mode and the channels are displayed, it can be seen that the fingerprint is most dominant in the green channel and least dominant in the red channel.
c. This offers the chance for a channel subtraction to at least partially isolate the fingerprint detail and remove some of the obstructive background (Figure 5.14).
d. When the red and green channels are subtracted, the result is an improvement in the content and clarity of the ridge detail, but

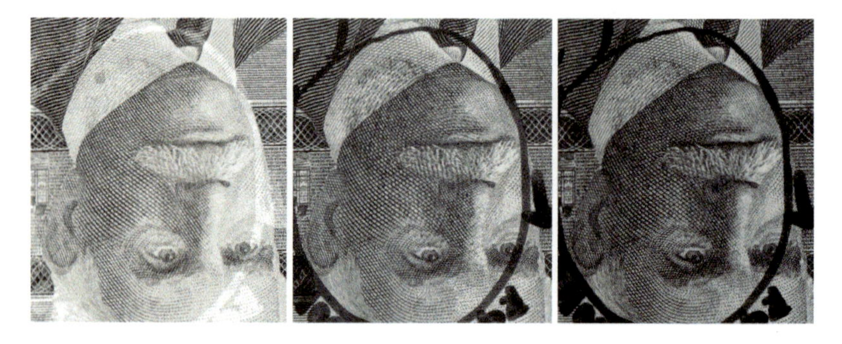

Figure 5.14 Left to Right: Red channel, green channel, blue channel. Fingerprint is virtually absent from the red channel and strongest in the green channel.

Figure 5.15 Result of subtracting the red channel from the green channel.

there is still significant obstruction due to the engraving on the bill. The inks used in the printing of the bill are complex in color composition and are not conveniently confined to one color channel. This makes them difficult to subtract completely (Figure 5.15).

e. Next, the image is opened in Lab color mode (see Chapter 3). The image looks no different, but there has been a very different distribution of the color data, as can be seen in the component channels (Figure 5.16).

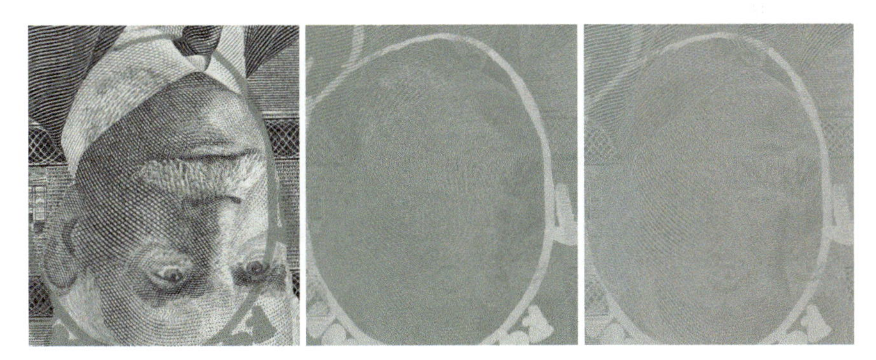

Figure 5.16 LAB color channels for Figure 5.13 (lightness, "a," and "b"). Fingerprint detail is easily visualized in the "a" channel, and not visible in the "b," although substrate pattern is visible.

Figure 5.17 Result of "b" channel subtracted from the "a" channel.

f. When the "b" channel is subtracted from the "a" channel, the clarity of the fingerprint is slightly better than the result of the RGB subtraction, but obstruction from the engraving pattern in the background still hides ridge detail. This image is saved and reopened in Image-Pro Premier (Figure 5.17).

Using FFT in Image-Pro Premier

The opening screen for Image-Pro Premier is depicted in the section that follows (Figure 5.18).

1. The first step in using FFT is to open the file and examine the obstructing pattern that one wishes to eliminate. If it is uniform and unchanging across the entire area occupied by the fingerprint, there is no need for neighborhood editing, and it may be conducted in one step on the entire image.
   a. Select *Process,* and click on *Forward* in the *FFT icon.* This transforms the image data to the periodic or frequency display (Figure 5.19).
   b. Select *Adjust* and the BL WL.

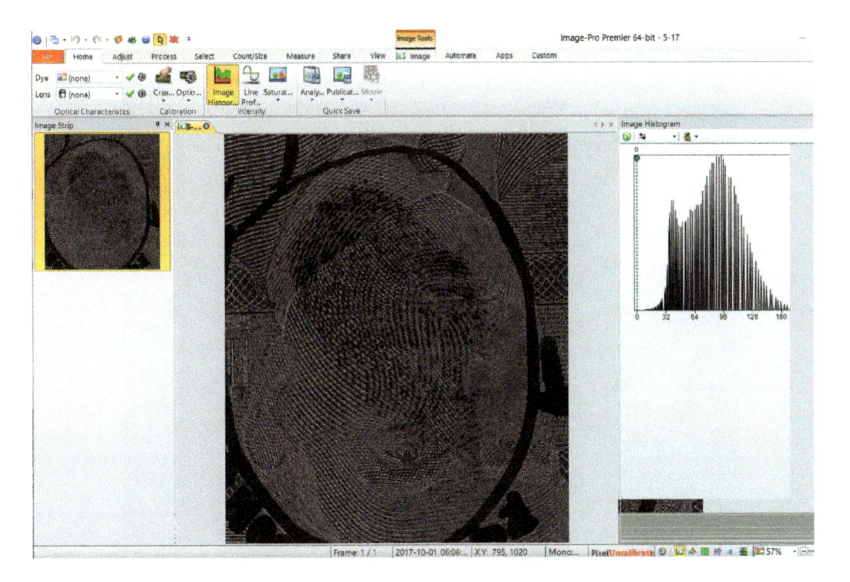

Figure 5.18 Image-Pro Premier, opening screen—Figure 5.13 open.

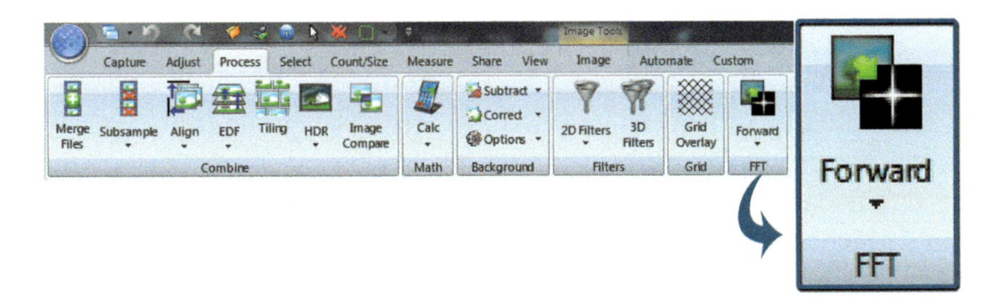

Figure 5.19 Detail of screen showing the FFT button.

c. Set the black level at "0" and the white level at "1."
d. Click on the periodic display, and any signature spikes of repetitive pattern are visible.
e. When the image data are organized as a function of frequency, the display no longer looks like a conventional image. It is a predominantly black rectangle divided into four quadrants.
f. A bright star or cross is usually seen in the center of the periodic display. Signature spikes associated with any repetitive pattern are symmetrical and are "mirrored" in the diagonal quadrant. In Image-Pro Premier, editing done in one quadrant automatically is applied to the opposite diagonal quadrant.
g. As illustrated previously, an image that does not contain repeating patterns displays virtually nothing in the quadrants when converted to the periodic domain.

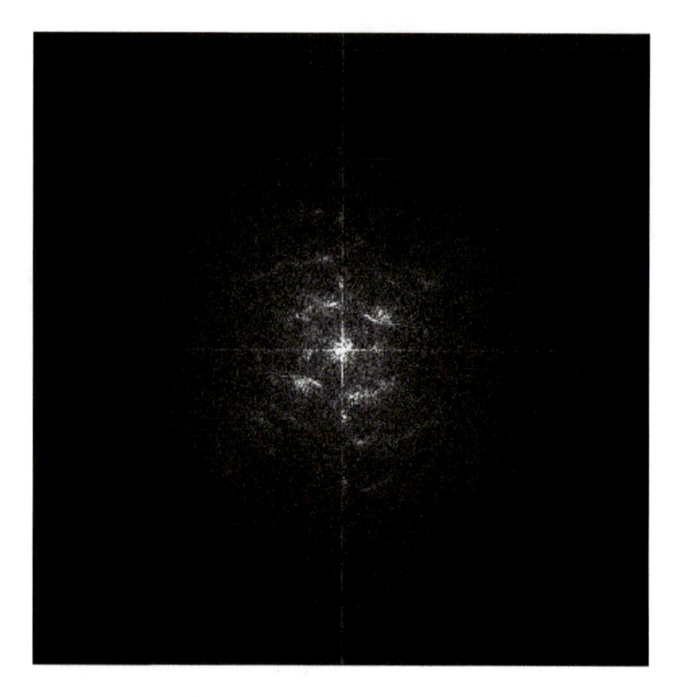

Figure 5.20 Periodic display of entire image in Figure 5.13.

h. Unlike the other images with obstructive backgrounds previously featured in this chapter, the engraving elements on the surface of the banknote vary considerably in direction, strength, and frequency from one part of the image to another. The upshot of this is a more complex signature when the entire image is viewed in the periodic display. It is difficult to interpret and, consequently, much more difficult to edit (Figure 5.20).

Image-Pro Premier allows for the transform of the image, one selected area at a time, as indicated by the green rectangle in Figure 5.21.

The signatures of the fingerprint and obstructive pattern can be more easily recognized. The fingerprint data are seen in the blue oval below, while the signature spikes of the background (noise) are pink. They are also seen to be in different locations of the periodic display from the fingerprint, facilitating the editing process.

When the editing in this area is completed, another area is selected, ensuring that it overlaps the first (Figure 5.22).

The direction of the ridges in this new area is at a different angle than the ridges in the first area selected, and the corresponding signature is in a different part of the periodic display. The spikes representing the obstructive pattern (pink) are removed, and the process is repeated until all background

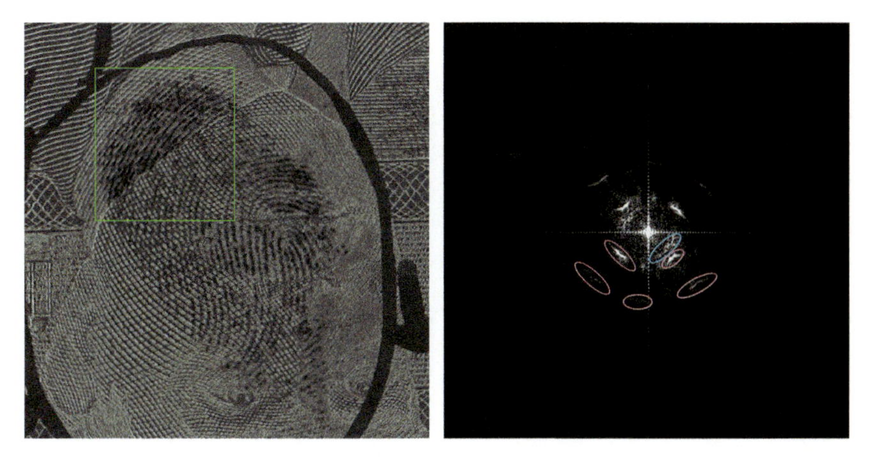

Figure 5.21 First selected area of Figure 5.13 and the corresponding periodic display.

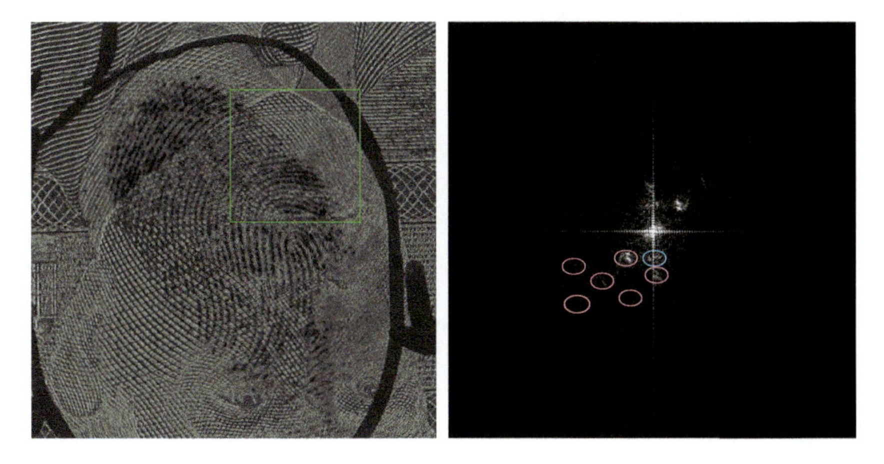

Figure 5.22 Second selected area of Figure 5.13 and corresponding periodic display.

pattern in the area of the image occupied by the fingerprint has been removed.

A comparison of the edited image with the initial image (see Figure 5.23) demonstrates that a significant improvement in signal-to-noise ratio has been achieved, and that the fingerprint detail is much easier to analyze and evaluate than in the original image. Again, it must be stressed that the fingerprint data was not altered during the FFT operations. This is a removal of noise to reveal the fingerprint detail that was always present, accomplished with a sequence of channel subtraction and FFT editing.

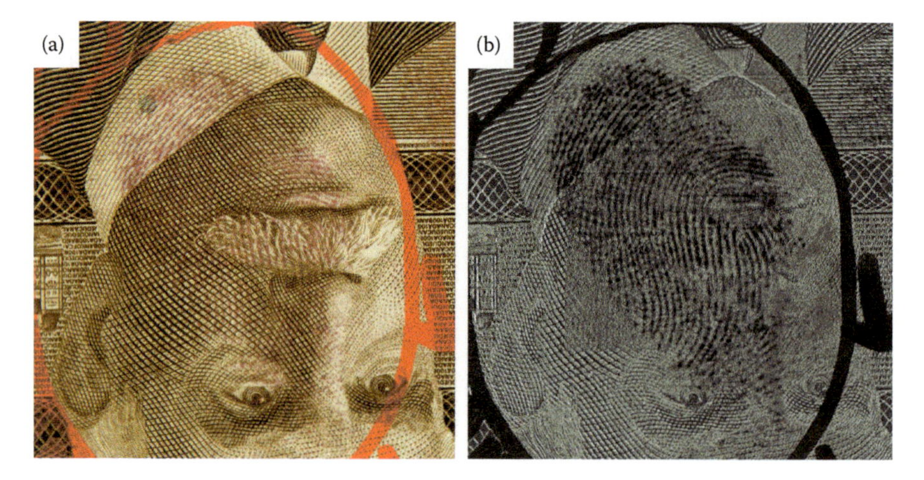

Figure 5.23 (a) Original Figure 5.13. (b) Results of editing after subtraction and FFT pattern removal.

Image Optimization and Sequence of Actions

1. Photoshop
   a. Image opened as RGB file
   b. Conversion to Lab Color Mode
   c. Channel subtraction—"a" and "b" channels
   d. Resulting image saved and opened in Image-Pro Premier
2. Image-Pro Premier
   a. FFT
   b. Editing of periodic noise by neighborhood
   c. Resulting image saved

After channel subtraction, this image featured a stronger fingerprint signal, but it is still obstructed by the engraving pattern (noise), which exhibits a series of line functions that are parallel but changing in direction, strength, and distance apart. When the entire image is converted to the periodic display, the noise signature (white spikes) is quite complex and difficult to edit. The editing process was greatly simplified by a software option that allows the transform and editing of an image, one area of interest at a time.

Example 2 —Narrow Band Filter Photography Followed by FFT

The image that follows features an untreated fingerprint on a train ticket, revealed by a TracER laser, 532 nm, and photographed with a Coherent orange barrier filter in combination with an FF-1.0 narrow band filter

Figure 5.24 Untreated fingerprint on a train ticket, revealed by a Coherent TracER laser, and photographed with a combination of a Coherent barrier filter, and an FF-1.0 narrow band filter.

(Arrowhead Forensics) with a bandwidth of 10 nm and a peak transmission of 560 nm (Figure 5.24).

When the image is converted to grayscale, inverted, and adjusted for contrast in Levels, it is immediately evident that the printed dot pattern on the ticket is obstructing the ridge detail to a degree that renders it unsuitable for analysis, comparison and evaluation (Figure 5.25).

The obstructive dot pattern is uniform in that it displays little or no variation in the area occupied by the fingerprint. This means that its corresponding signature in the periodic display will be simple to locate and easy to edit (Figure 5.26).

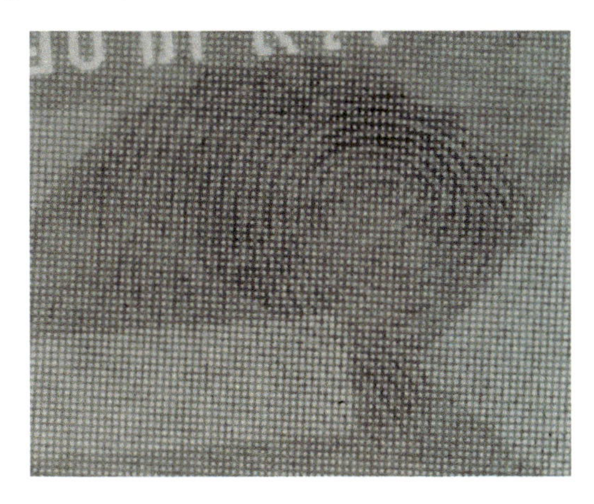

Figure 5.25 Grayscale image of Figure 5.24, revealing obstructive dot pattern.

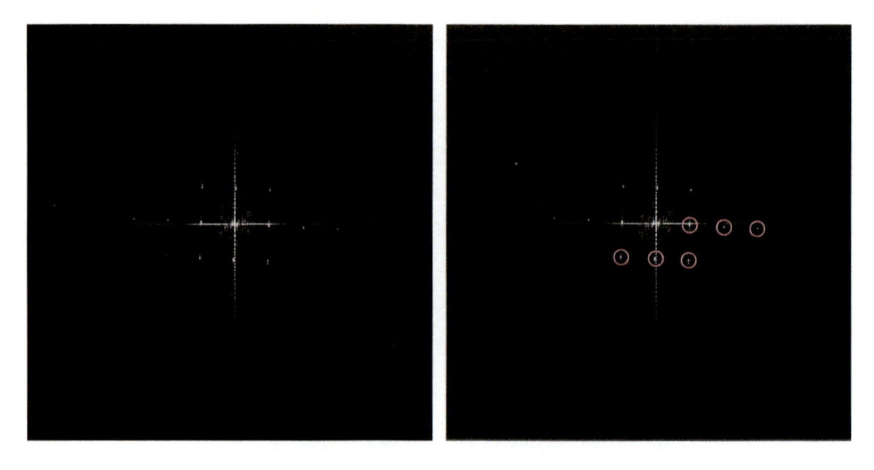

Figure 5.26 Periodic display of entire image (Figure 5.24), revealing (a) noise signature and (b) areas edited.

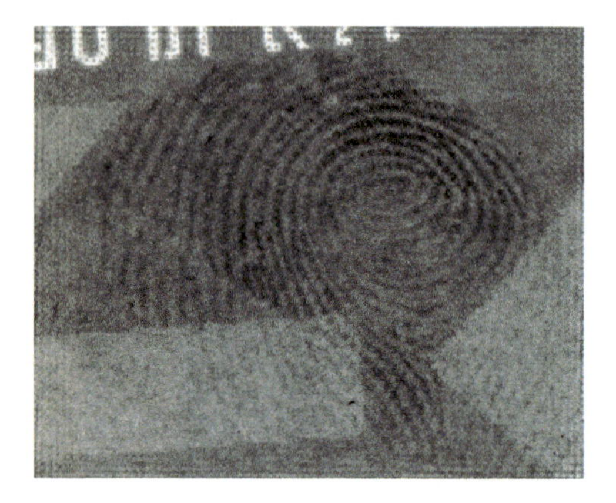

Figure 5.27 Figure 5.24 after FFT editing, revealing clear and continuous ridge detail.

The image was opened in Image-Pro Premier, and the region containing ridge detail was edited in fast Fourier transform, which resulted in the suppression of the dot pattern. Contrast adjustment was completed in Photoshop using Levels, with the result in Figure 5.27.

Image Optimization and Sequence of Actions

1. Photoshop
   a. Image opened as RGB file
   b. Conversion to grayscale file
   c. Invert
   d. Contrast adjusted in Levels
   e. Image saved

2. Image-Pro Premier
   a. Image opened as grayscale file
   b. Transformed in FFT to periodic display
   c. Editing of periodic noise
   d. Resulting image saved

Non-Fingerprint Applications

FFT can be used to optimize other types of obstructed evidence images, not just those containing fingerprints. It can also effectively reduce the effect of grain when wood is the substrate or the parallel lines inherent to some metal surfaces such as brushed aluminum. Several paintings from the early part of the twentieth century were examined by laser in efforts to find evidence establishing their provenance. A stamp impression was observed on a piece of the wooden frame of a painting. It was virtually invisible in room light (Figure 5.28).

When the frame was examined by laser and photographed using both a Coherent orange barrier and an Arrowhead FF-1.0 narrow band filter, the image of the stamp was much stronger (Figure 5.29).

The image was converted to grayscale. The woodgrain was assertive and obstructed the detail in the stamp. Any attempt at this stage to adjust contrast would result in the woodgrain becoming even more obstructive (Figure 5.30).

Editing in FFT was applied selectively to the area containing the stamp to reduce the woodgrain pattern (Figure 5.31).

Figure 5.28 Stamp impression on a wooden frame of a painting, in room light, revealed by a Coherent TracER laser. (Courtesy of Dana Rosenthal, art collector and appraiser.)

Figure 5.29 Stamp impression revealed by a Coherent TracER laser, photographed with combination of a Coherent barrier filter and FF-1.0 narrow band filter. (Courtesy of Dana Rosenthal, art collector and appraiser.)

Figure 5.30 Grayscale image of Figure 5.29, revealing obstruction of stamp by woodgrain. (Courtesy of Dana Rosenthal, art collector and appraiser.)

Figure 5.31 Figure 5.29 after editing, showing suppression of woodgrain. (Courtesy of Dana Rosenthal, art collector and appraiser.)

Summary

There are occasions when latent fingerprint detail is obscured by patterns in the substrate, and other digital procedures may not be successful in reducing or eliminating the obstruction.

Learning to read and interpret the detail in the periodic display is like learning a new language. Initially, it is unfamiliar and perhaps foreign in concept, but practice and repetition leads to a high comfort level in its use.

Editing a repetitive pattern in the Fourier domain is entirely defendable. The signature data associated with fingerprint detail are identified and left untouched.

The process is applied to a broad area without the anticipation of a specific desired outcome, other than the removal of periodic noise. It is entirely neutral and objective.

Complex pattern elements, such as the one observed on the paper currency, can change in direction, frequency, size, and gray value. When converted to the periodic display, the resulting signature can be intricate and difficult to read. Editing the image one area at a time can result in much simpler periodic displays, where the signature of the disruptive noise can be more easily identified and removed.

FFT can be used to remove or reduce obstruction caused not only by printed mechanical pattern elements but also more organic and variable interference such as that produced by woodgrain.

The goal in the application of FFT, as with all enhancement procedures, is to optimize the signal-to-noise ratio in evidence images.

Review Questions

1. What is meant by the periodic domain and display relating to an image?
2. Is a fingerprint a repeating pattern?
3. What is meant by an inverse transform?
4. Where should one look in the frequency display for the signature of a parallel line pattern that runs in a diagonal direction in the image from upper left to lower right?
5. What four descriptors determine the location and strength of the signature for repeating pattern in the frequency/periodic display?

References

1. C. Joseph, Unbaking a cake—Fourier series/transforms, available at https://www.youtube.com/watch?v=Qm84XIoTy0s

2. R. D. Olsen, Sr., and Charles C Thomas, Publisher, *Scott's Fingerprint Mechanics*, pp. 429–435, 1978.
3. W. J. Watling, Using the FFT in forensic digital image enhancement, *Journal of Forensic Identification*, 43(6), pp. 573–583, 1993.
4. E. Kaymaz and S. Mitra, A novel approach to Fourier spectral enhancement of laser-luminescent fingerprint images, *Journal of Forensic Sciences*, 38(3), 1993.
5. B. E. Dalrymple and T. Menzies, Computer enhancement of evidence through background noise suppression, *Journal of Forensic Sciences*, 39(2), pp. 537–546, 1994.
6. Erik Berg, personal communication, 2017.

6 Contrast Adjustment Techniques

> When one door of happiness closes, another opens; but often we look so long at the closed door that we do not see the one which has been opened for us.
>
> **Helen Keller**

This chapter covers the commonly used contrast adjustment tools that are often the last step of the optimization process. There is a brief introduction into the Photoshop workspace, a review of the tools relevant to forensic image processing and where to find them, followed by contrast adjustment functions and how they work.

Toolbar

The toolbar in Photoshop runs vertically down the left side of the desktop. Some tools have a tiny black arrow in the bottom right corner of the icon. This indicates a rollout menu, with more options underneath that can be accessed by clicking and holding the mouse button down. Many tools are used for artistic purposes and are not a concern for forensic use, so here are a few of the tools we use and their functions (as listed in Figure 6.1).

Move Tool (V)

The move tool is used to move layers or objects within a layered file.

Selection Tools

While it is desirable to make global adjustments to digital images as much as possible, at times it is necessary to select portions of the image and only adjust a specific area. There are many options in Photoshop for creating selections within an image, plus a few considerations of which you should be aware. Some selection tools are classified as marquee tools, lasso tools, and others fit into the wand tool category.

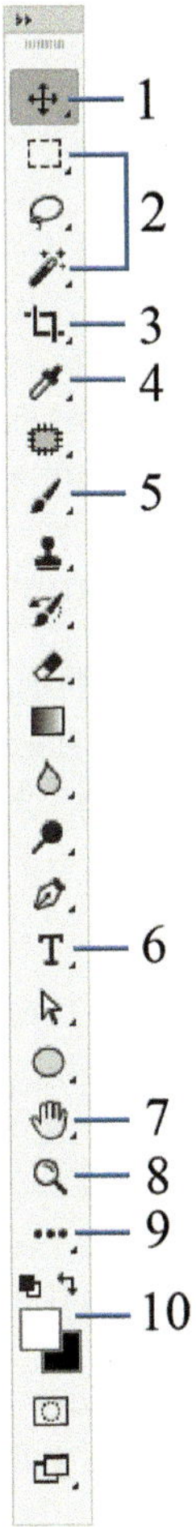

Figure 6.1 Toolbar. (1) Move tool. (2) Selection tools. (3) Crop tool. (4) Eyedropper (ruler). (5) Brushes. (6) Text tool. (7) Hand tool. (8) Zoom tool. (9) Edit toolbar. (10) Foreground and background colors.

Marquee Tools

The first selection tool in the toolbar is directly under the move tool—the marquee tools. The marquee selection options are (Figure 6.2):

- Rectangle Marquee
- Elliptical Marquee
- Single Row and Column Marquees (vertical and horizontal)

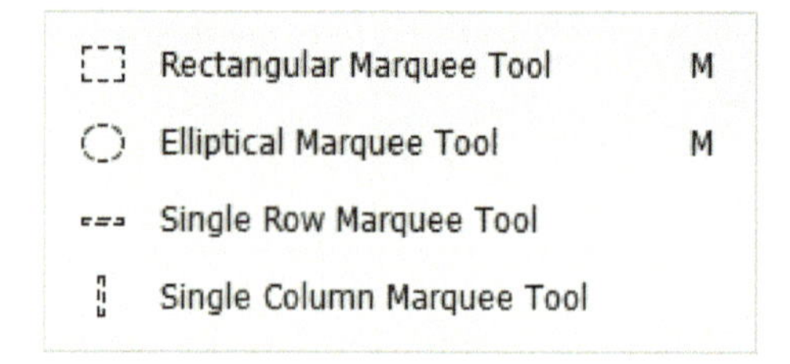

Figure 6.2 The marquee rollout menu.

When activated, just click and drag out the marquee size needed. Holding the shift key while dragging the rectangle or elliptical creates a perfect square or circle respectively.

Lasso Tools (Figure 6.3)

The Lasso

The Lasso tool allows free-form selections to be made of an area within the image. Click and hold the mouse to draw around an area of interest. When released, the endpoints will connect. The lasso tool may be used for making freehand selections around dark or light areas of interest.

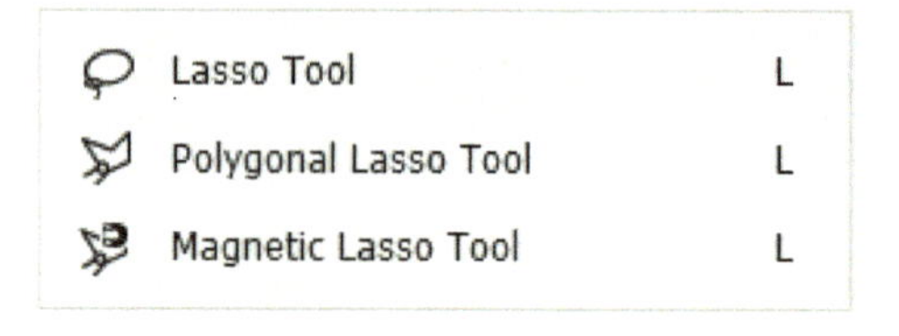

Figure 6.3 The lasso rollout menu.

The Polygonal Lasso

This selection tool draws in straight lines. Click to set a starting point (holding the shift key limits the lines to horizontal, vertical or 45 degrees). To finish the selection, click on the starting point again to close the area, or double-click to close the points.

The Magnetic Lasso

This tool automatically snaps to the edges of defined areas of contrast in an image. Click to start, and begin tracing the defined area of interest—control points are placed along the edges as the pointer-cursor travels. The sensitivity of the magnetic function of the tool can be customized in the options bar across the top of the desktop:

Width—Select a pixel value for width. The magnetic lasso tool detects edges within the specified distance from the pointer.

Contrast—Select a percentage from 1 to 100. The higher the percentage, the more contrast required between tonal values to define an edge.
Frequency—Enter a value between 0 and 100. This value affects the rate at which control points are fastened to the edge as the line is drawn. A higher value increases the rate at which the points are placed.

Wand Tools

Magic Wand

The Magic Wand Tool (Figure 6.4) allows you to select an area of similar color and tonal value (based on the tolerance level you set), with one click within the area of interest.

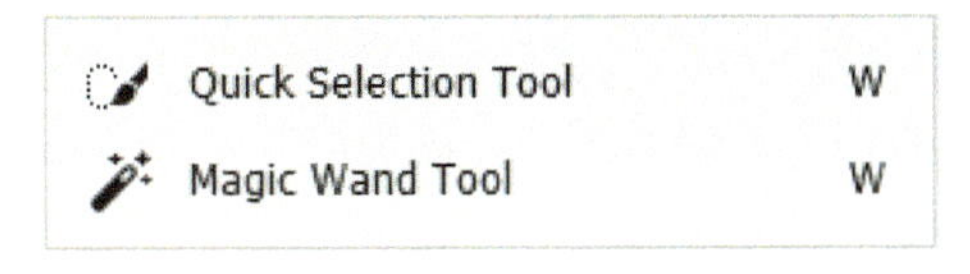

Figure 6.4 The magic wand rollout menu.

The extra tool options (found in the tool options bar just under the menu bar):

Sample size—3 × 3 or 5 × 5 recommended.
Tolerance level—The tonal/color range of pixels to be selected. Enter a value between 0 and 255. A low number limits the range of tonal values selected, and a higher number broadens the range of tonal values selected.
Contiguous—Selects areas of tonal values/colors next to the area selected. Otherwise, all pixels in the entire image using the same tonal values/colors are selected.
Sample all layers—If checked, it selects tonal values/colors using data from all layers.

Quick Selection

Use the cursor to follow defined edges in the image. After making the *new* selection, any *painting* done with the Quick Selection Tool automatically adds to the selection. The extra tool options (again, found in the tool options bar, under the menu bar):

Brush size—Enter the pixel diameter desired for brush size.
Sample all layers—If checked, it selects tonal values/colors using data from all layers.
Auto enhance—Smooths the selection boundary.
One caveat of the quick selection tool—*It is not compatible with Actions when recording processing steps, as it functions much like a brush.*

Crop Tool (C)

Crop any rectangle or square section of an image.

Eyedropper Tool (I)

May be used to read the intensity and/or color values of any area within an image. More importantly, the ruler tool is found in this rollout menu (required to calibrate images 1 to 1 for AFIS). Click and hold the tool to see all options in the rollout menu.

Brushes Tool (B)

May be used with adjustment layers with a layer mask.

Text Tool (T)

May be used to add text to an image.

Hand Tool (H)

When zoomed into an image, the hand tool may be used to move the image around within the zoom window. Holding spacebar down may also be used to temporarily switch to the hand tool from any other active tool; letting go of the spacebar returns the cursor to the active tool.

Zoom Tool (Z)

When selected, zoom into an image; hold the Alt key down to zoom out again.

Edit Toolbar

(Also found under Menu Bar > Edit > Toolbar) Customize the toolbar and group tools as desired:

- Drag and drop tools and/or groups to re-organize the toolbar.
- Move unused, or unnecessary tools to Extra Tools.
- To access Extra Tools, click and hold icon at the bottom of the toolbar.
- Save a custom toolbar by clicking *Save Preset.*
- Open a previously saved toolbar by clicking "Load Preset."
- Click "Restore Defaults" to restore the default toolbar.
- "Clear Tools" to move all tools to Extra Tools.
- Foreground and Background colors.

Tool Options Bar

Once a tool has been activated from the toolbar, the options for that specific tool are displayed across the top of the workspace directly beneath the menu bar. Some options are common to all selection tools, and some are unique to a specific tool.

Figure 6.5 The Options Bar in Photoshop.

Tool Icon

The first part of the tool options bar, beginning from the left, is the icon representing the activated tool. In Figure 6.5, it is the magnetic lasso tool.

Selection Options

The four selection options shown in Figure 6.5 are in order from left to right:

- New selection
- Add to selection
- Subtract from selection
- Intersect with selection

Feather—There is a Feather option box to indicate the number of pixels used to blend the selected area of interest into the image. It is usually preferable to leave this at 0 and instead set the feather values in the Select and Mask dialogue box.

Anti-alias—Check this option to create a smoother edge selection.

Width, Contrast, and Frequency—These options refer to the magnetic lasso tool options (see magnetic lasso description).

Select and Mask—Open the Select and Mask dialogue box.

About Feather and the Select and Mask Dialogue Box

When making a selection, particularly if it's to be used for later analysis, it is important to feather the selection before applying any adjustments. What does that mean? It means that there won't be a hard edge delineating the part of the image that has been adjusted for brightness and contrast from the part that has not; applying a feather to a selection softens the effect of the contrast adjustment over a wider range so that it blends softly without creating any distracting edges. In most cases, it is preferable to feather generously when making a selection. How much of a feather value is necessary? That depends on the resolution of the image. The higher the resolution of the image, the larger the feather value required to blend the effect. The number is not important, provided that the effect is blended well to avoid harsh edges—for instance, my camera's chip captures at a resolution of 36.3 megapixels. After calibrating a copy of my original, any selection I make is feathered by anywhere from 70 pixels to 370 pixels.

It is recommended to set your feather values in the *Select and Mask* dialogue box rather than in the tool options bar. Setting a feather value in the tool options bar saves it there until it is manually changed; this is a nuisance if it applies this feather to a selection painstakingly made at a later date, requiring the user to start over if the feather was not desired. There are several ways to get to the Select and Mask dialogue box:

1. It is the last option offered in the tool options bar when you have a selection tool active.
2. Right-click within your selection, and choose *Select and Mask* or *Feather*.
3. Go to Select > Select and Mask.

A Look at the Select and Mask Dialogue Box

You may make a selection and then open the Select and Mask dialogue box, or you may go directly into the Select and Mask dialogue box and make the selection there (Figure 6.6).

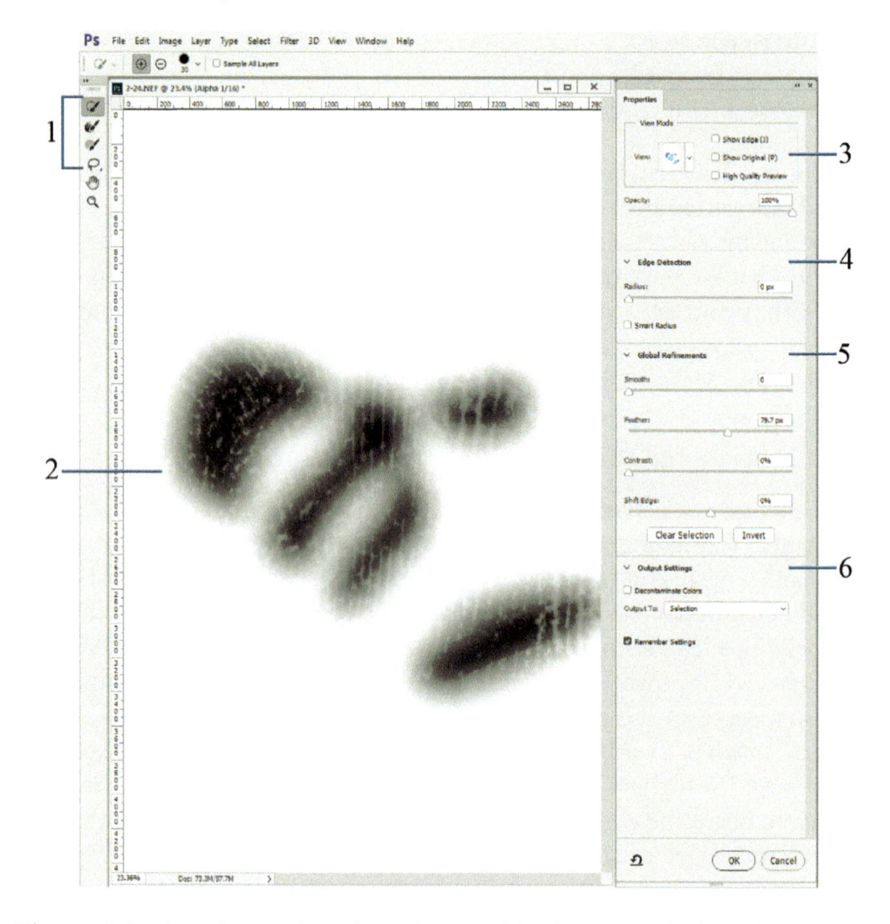

Figure 6.6 The Select and Mask workspace. (1) Selection tools. (2) Selection with feather preview in workspace—View mode: On white. (3) View mode options. (4) Edge detection options. (5) Global refinements. (6) Output settings.

The Tools within Select and Mask

The tools are listed on the left-hand side of the workspace—they are listed here in order from top to bottom:

Quick Selection Tool The quick selection brush may be dragged across an area of similar values that you wish to select. This smart tool searches and finds contrast edges within the area you drag. *Note: This tool is not compatible with Actions for recording image processing steps.*

Refine Edge Brush Tool Adjusts the area of edge refinement. Brush over areas of fine details to include in selection. The brush tool is used after this step to clean up the selection.

Brush Tool With a rough selection already made, areas of the selection may be painted with the brush tool in Add mode (add to selection), or in Subtract mode (subtract from selection).

Lasso Tool Draw around the area of interest to be adjusted.

Hand Tool Use (click-drag) to navigate within the image.

Zoom Tool Zoom in or out of image.

The Options Bar within Select and Mask

As in classic Photoshop, the options bar runs across the top of the workspace and gives you options for the tool selected.

Add and Subtract Just as the names would imply, use these options to add or subtract from an existing selection using a brush. Adjust the brush to the desired size.

Sample All Layers When this option is checked, tools that automatically search for contrast edges to assist with making a selection, such as the quick selection tool, create a selection based on all layers.

The Properties Panel within Select and Mask

The Properties panel on the right side of the workspace offers many setting options to assist with refining a selection.

View Mode Settings The View Mode settings allow you to choose how you wish to view your selection. Consider the following options:

- Onion Skin—Animation-style onion skin scheme.
- Marching Ants—Marching ants, as in classic Photoshop.
- Overlay—Selection is visualized over a transparent overlay of color.
- On Black—Selection is visualized over a black background.
- On White—Selection is visualized over a white background.
- Black & White—Selection is visualized as a black and white mask.
- On Layers—Preview of selection displayed using all layers underneath.

**Press F to cycle through the View Mode menu.*

Show Edge—Shows the area of refinement in a selection.

Show Original—Shows the original selection.

High Quality Preview—View in high-quality preview (software response may slow).

Transparency/Opacity—Adjusts opacity of selection mask.

Edge Detection Setting

Radius—A small radius defines sharper edges, while a large radius defines softer edges.

Smart Radius—This tool allows for variable width refinement around the edge of a selection.

Global Refinement Settings

Smooth—Subtly smooths the edges of a selection.

Feather—Blurs and blends the transition between the selection and the surrounding area.

Contrast—Increasing the contrast of a selection makes the soft edges of a selection sharper.

Shift Edge—Adjusts a selection to contract inward (smaller) with a negative value or to expand outward (larger) with a positive value.

Output Settings

Decontaminate Colors—Replaces a soft color fringe (multiple layers) with colors sampled from fully selected pixels nearby (must output to a new layer or document).

Output to—A drop-down menu is available for output settings. Default setting is Selection (other options: Layer Mask, New Layer, New Layer with Layer Mask, New Document, New Document with Layer Mask).

The many tools and options in this dialogue box may seem daunting, but in terms of fingerprint adjustment, use a lasso, feather it, and click OK. Many of the other options are for the very complex selections required for artwork. Some of those may be used for changing backgrounds on mugshots submitted by an outside agency. Not all police agencies use the same color background when capturing mugshot images; when creating a lineup, it is necessary to place the individual on the same background as the other images in our database so that they don't stand out. That's where some of the edge refinement tools come in handy in preserving hair wisps for a realistic selection.

Point Sample

There are many tools that may require the "sampling" of tonal values or colors within an image to perform certain functions (curves, levels). Whether the eyedropper tool is selected from the toolbar or from within the curves dialogue box, it takes a reading of the tonal values within the image wherever the eyedropper is clicked.

The Photoshop default setting for the eyedropper tool in the tool options bar is "point sample." This means that wherever you click in the image, a sample is taken from that one, exact pixel. If the point sample ends up being taken of an errant bit of noise or color aberration, any adjustments made based on that value would be inaccurate. It is recommended that you change this setting to 3 × 3 average, or 5 × 5 average, wherein the point sample is averaged with the nearest surrounding 3 pixels if set to 3 × 3 average (or 5, if set to 5 × 5 average). This renders a more accurate sample of the area, allowing for small errors such as noise to be blended with its surroundings. To set sample size, activate the eyedropper tool from the toolbar (located underneath the crop tool in the first toolbar segment), and in the tool options bar, set sample size to 3 × 3 average. This setting will remain at 3 × 3 until it is manually changed.

Contrast Adjustments

All contrast adjustment tools have the potential to clip data. This may seem an alarming statement, and it is, but it does not mean that contrast tools must be avoided carte blanche. It *does* mean that we must be cognizant of how these tools function and what is happening to the image with more understanding than the average user. With training, the discerning forensic processor will understand the effects of contrast adjustments to the image and be aware of what parts of the image detail are being sacrificed, if any. We are not in the business of creating detail. We are in the business of *optimizing the signal-to-noise ratio*, meaning that we can only make an effort to clarify image information that is already present in the digital image, while suppressing any noise that may be visually distracting to the greatest degree possible.

One common scenario that is extremely frustrating to face is a photograph of an impression dusted with black powder on a metallic gray car; the lift is poor, and image processing is required. Having only the photo to go on and no color in either the ridge detail or the substrate to help facilitate some contrast separation, there is little to nothing that can be done. Any boost in contrast makes the metallic flecks in the paint become monstrous entities that obscure everything. Many images can be improved using the techniques in this book, but unfortunately, there are some that cannot be helped. If, however, the photographer takes a thoughtful, *proactive* approach to this problem at the scene, one of two things might be done that could render a usable fingerprint rather than a print of no value:

- Dust fingerprint impressions with colored powder so that some separation might be possible using color channel blending.
- Photograph fingerprint impressions with a tripod, then clean the substrate and photograph the background (subtraction).

Be prepared to face the inevitability that some images may not be salvageable, either because the signal can't be discerned from the substrate noise, or because it simply isn't there in the first place.

We have waited until Chapter 6 to address contrast adjustment tools because there *is* a pecking order to image processing steps. This book attempts to address forensic image processing strategies as linearly as possible, introducing each skill in the order in which it should be applied.

There are four grossly generalized steps to forensic image processing:

1. Adjust RAW image in Camera Raw (or other processing software) to best visualize detail in shadow, highlight, and midtone areas of the impression.
2. Prepare and size for AFIS (1:1).
3. Minimize distracting background patterns (image subtraction, FFT, or channel blending).
4. Adjust contrast.

In Chapter 3, *Apply Image* and *Calculations* were introduced as a possible way to minimize noise, strengthen signal, and/or boost contrast and brightness. For example, the extremely compressed dynamic range of the “a” and “b” channels in LAB color mode are often greatly improved by applying them to themselves in overlay mode, significantly increasing the contrast without clipping any highlights or shadows (because there aren't any), but those are very general techniques that help get us into the ballpark, and now we require tools that offer a little more precision and impact. It's time to review the rest of the contrast adjustment tools in Photoshop.

Levels

The levels dialogue box can be found by going to the Menu bar > Image > Adjust > Levels (keyboard shortcut: Ctrl—L).

The levels *histogram* illustrates how pixels in an image are distributed by graphing the number of pixels at each intensity level from 0 (black) to 255 (white), and it allows you to adjust the tonal range of shadows, midtones, and highlights.

In an average-key scene capture with good exposure, the histogram may look similar to Figure 6.7. Most of the pixels are heavily weighted in the middle of the graph, with a few highlights and shadows plotted at the white and black ends. Low contrast images may not have any values plotted at the far left *or* right of the histogram, meaning there are no shadows (blacks) *or* highlights (whites) in the image. Every image needs some black and white to

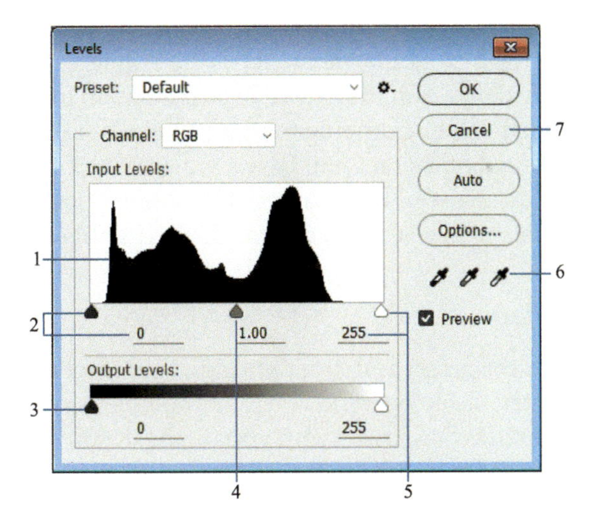

Figure 6.7 Levels dialogue box. (1) The histogram graph. (2) Black point slider and its tonal value. (3) Output level sliders. (4) Gamma slider. (5) White point slider and its tonal value. (6) Eyedroppers. (7) Cancel button which toggles to reset when the Alt key is held down.

represent the shadows and highlights to make the image look natural, and most digital images require some adjustment to achieve this.

The Histogram Graph

The tonal values of the pixels are plotted on a histogram graph—all 256 shades of gray. Black pixels are plotted at the far left at 0, and white pixels are plotted at the far right at 255, and all other shades of gray are plotted in between, with the midtones in the 128-tonal value range.

Black, White, and Gamma Sliders

There are three triangle-shaped sliders located directly under the histogram that can be used to make tonal adjustments to the image. The black point slider (far left) sets a new black point for the image as you drag it to the right, darkening the image, especially in the shadows. The further you drag the slider, the darker values are now defined as black rather than dark gray. The white point slider works the same way—drag the white point slider to the left, and the tonal values within that range become pure white, lightening the image overall, especially in the highlights. Any pixel values that fall between the new black point and 0 will be clipped, and any values that fall between the new white point and 255 will be clipped (more on clipping later). The gamma slider can be adjusted to lighten or darken the midtones in the image without affecting the blacks and whites. Once the black and white points in the histogram are set, all other brightness values are redistributed or *stretched* between them. As these brightness values are stretched, gaps appear along the histogram where there is no brightness representation; overstretching may then result in noticeable posterization of the image, with ugly patches of tonal values clumping together in blocks rather than a tonal gradation of shading. For this reason, Levels is a great tool for tweaking the contrast to a modest degree, while saving the more robust contrast adjustments for other tools such as Curves and Shadows/Highlights.

The Output Sliders

Found underneath the graph and input settings, they can be used to darken white tones or lighten black tones; however, this does not recover any clipped detail and offers no benefit to assist with visualizing detail.

Eyedroppers

There are three eyedroppers to the right of the histogram. When the black dropper is clicked on a shadow within an image, it sets those tonal values to black, and when you click on a light area in the image with the white dropper, it sets those values to white. The gray dropper may be clicked on a known neutral tone in a color image to correct color casts.

Auto and Options

There are Auto and Options buttons under the Cancel button; these have to do with allowing the software to make adjustments automatically. As forensic imaging experts, we do not require the use of these. Avoid using "autocorrect" controls to every degree possible; make your adjustments manually.

Tip: *When making adjustments in any dialogue box in Photoshop, you may reset your settings if they are undesirable by holding the Alt key down. The Cancel button will change to a Reset button.*

Clipping

When the black and white points are dragged beyond where values and potential detail are plotted on the graph, those values become *clipped*. Anything from the new black point (the value is displayed in a window under the shadows side of the histogram) to the 0 point, now becomes 0 (black), and any tonal values existing between the new white point to 255 now becomes white. Any detail that existed there is permanently lost (clipped). Pixels climbing up the left/black or right/white side of the graph indicate pure black or white and the possibility of *clipping*. Why do we use the term "possibly clipped?" It takes human interpretation to determine if important detail has been clipped in a given image. An icy landscape featuring the sun shining brightly through the clouds most assuredly has pixels climbing the white side of the graph. The hot spot of the sun and the tiny spectral highlights glinting off the icicles do not carry detail, and we would expect those values to be plotted up the right side of the graph at 255. If the tonal values of the entire icicle have been clipped, then we have a problem, and the image has been either grossly overexposed, or someone has been too heavy handed with one of the contrast adjustment tools and lost vital image data. Remember, as forensic imaging professionals, we are working on copies of our originals, so if we have gone too far, we can always start over.

Making Levels Adjustments and Interpreting the Histogram

The ability to read a levels histogram not only helps avoid clipping signal detail while making adjustments, but it offers a significant benefit to photography at the capture stage. The histogram reveals if an image has been underexposed or overexposed at the point of capture. LCD screens are notorious for their unreliability in terms of judging exposure, color, or focus. Understanding what to expect from the histograms of high key images (light tonal values), low key images (dark tonal values), and average key images is important.

Figure 6.8 is an image of a fingerprint from a ninhydrin-developed document. It was converted to LAB color mode, and the "a" channel was applied to itself in overlay mode twice to increase the contrast, but it still has a very compressed dynamic range. An evaluation of the levels histogram reveals that there are no shadows or highlights in the image. Setting a black and white point in the histogram does much to improve the contrast, and we will be careful to avoid clipping fingerprint image information.

Tip: *It is possible to visualize which areas are being clipped for either shadows or highlights when setting the black and white points in Levels. Hold the Alt key down while adjusting the points, and the window previews the shadow areas or highlight areas that are being clipped as you drag. This works in RGB color mode and grayscale mode. It does not work on an alpha channel until it is converted to grayscale.*

The black point slider of Figure 6.9 has been dragged to the value of 83 (the base of the graph-mountain where most of the pixel values are concentrated)

Figure 6.8 Average-key image with normal exposure—Very low dynamic range (contrast). (Courtesy of Andrew Cholmondeley, York Regional Police. All rights reserved.)

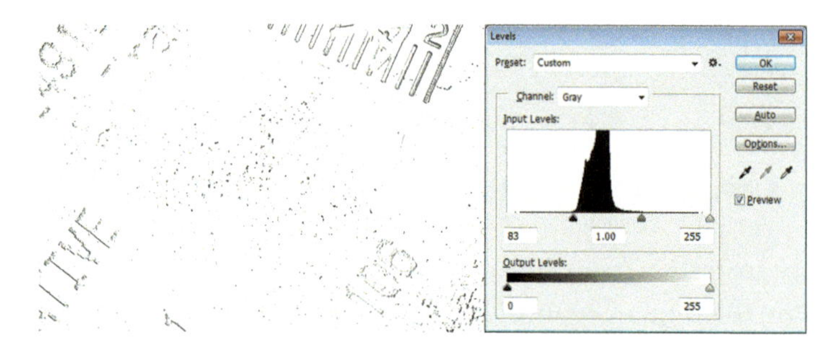

Figure 6.9 Shows where the shadows are being clipped in the image.

while holding down the Alt key. The image window is a visual preview of the shadows being clipped in the fingerprint ridge areas. Those areas of the ridge itself are now an area of black rather than shades of dark gray and black. Because there is evidence of clipping occurring in the ridge detail, the slider may be pulled back from that setting to avoid loss of detail in that area. When the Alt key/mouse is released, the image returns to the window with those settings.

The fingerprint has been improved by setting a black point, but it is still lacking a highlight. In the same levels dialogue box, before clicking OK, the highlight must be adjusted. Figure 6.10 illustrates where the highlights are being clipped as the white point is dragged to the value of 144 while holding the Alt key.

Figure 6.11 is the result after applying levels. The goal is not to make it look like a rolled black and white print but to make the ridges easier to visualize. Some clipping has occurred in the scale and the text *outside* the area of interest, and you can see those pixels climbing the black and white sides of the histogram. I find the clipping of those areas to be acceptable, as they are well away from the area of interest.

Figure 6.10 Shows where the highlights are being clipped in the image.

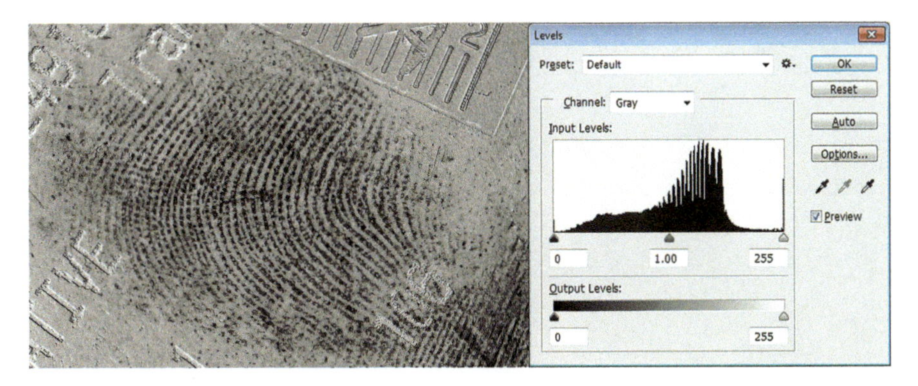

Figure 6.11 Fingerprint image adjusted with Levels. (Courtesy of Andrew Cholmondeley, York Regional Police. All rights reserved.)

Figure 6.12 An average-key image with a normal exposure.

Figure 6.12 shows the levels histogram of an average-key image with a normal exposure. Most of the pixels are plotted in the midtone range, tapering off near both ends of the histogram, indicating that there are no shadows or highlights. An adjustment to establish a definite shadow and highlight will go far to improve the depth of this image. The black point slider is dragged in to the value of 51. Any values between 0 and 51 are permanently changed to 0 and clipped. The white point is dragged in to the value of 200. Any values between 200 and 255 are permanently changed to 255 (white). Those values are also clipped. I find the clipping of the darkest shadows as well as the five tonal values in the highlights to be acceptable as they don't contain detail that I believe is important to the image.

In Figure 6.13, see the results of the levels adjustment and the new histogram.

Figure 6.14 is a very low-key image with a healthy histogram for that image. This is a picture of my son in a pool at night with the corresponding histogram. Many pixel values are plotted in the midtone range, representing

Figure 6.13 A subtle contrast and brightness adjustment made in Levels.

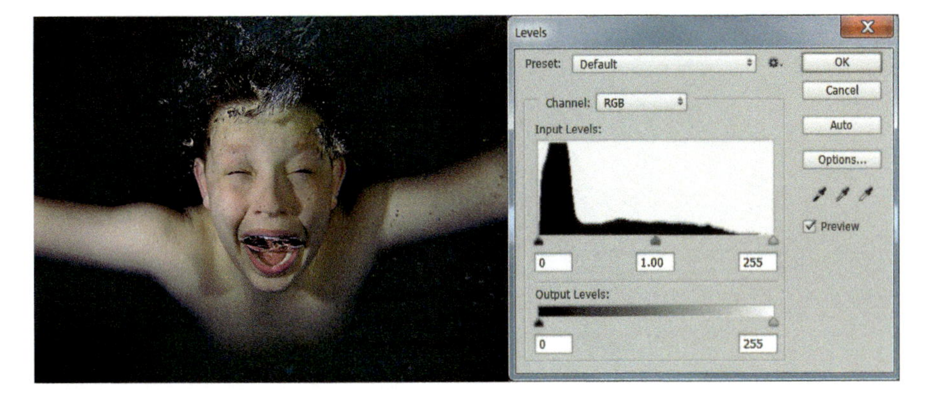

Figure 6.14 A dark, low-key image correctly exposed.

Figure 6.15 A high-key image normally exposed.

Sam's skin tones. A few pixel values are plotted at the light end, representing the highlights in the bubbles, but the lion's share of pixels have been plotted at the dark end of the graph, just short of black, representing the dark background. If the histogram looked like this on a sunny day in the pool, we would be alerted to a severely underexposed image and re-take the shot.

Conversely, Figure 6.15 is a high-key image of eggs in white bowl. Most of the pixel values are plotted on the light end of the graph, as expected, with some of the highlights near white.

Impression images for analysis, however, are not your average scene, nor do they present the same way in a levels histogram. Perfectly rolled fingerprints might be expected to be graphed as one large spike of pixel values plotted near 0 and another separate large spike near 255; yet impression evidence rarely presents itself that way. Figure 6.16 shows three different fingerprint images in grayscale, with their corresponding levels histograms to their right. The top image has been *manufactured* to exaggerate the black ridges and white furrows for illustration purposes only. There is a spike of pixels

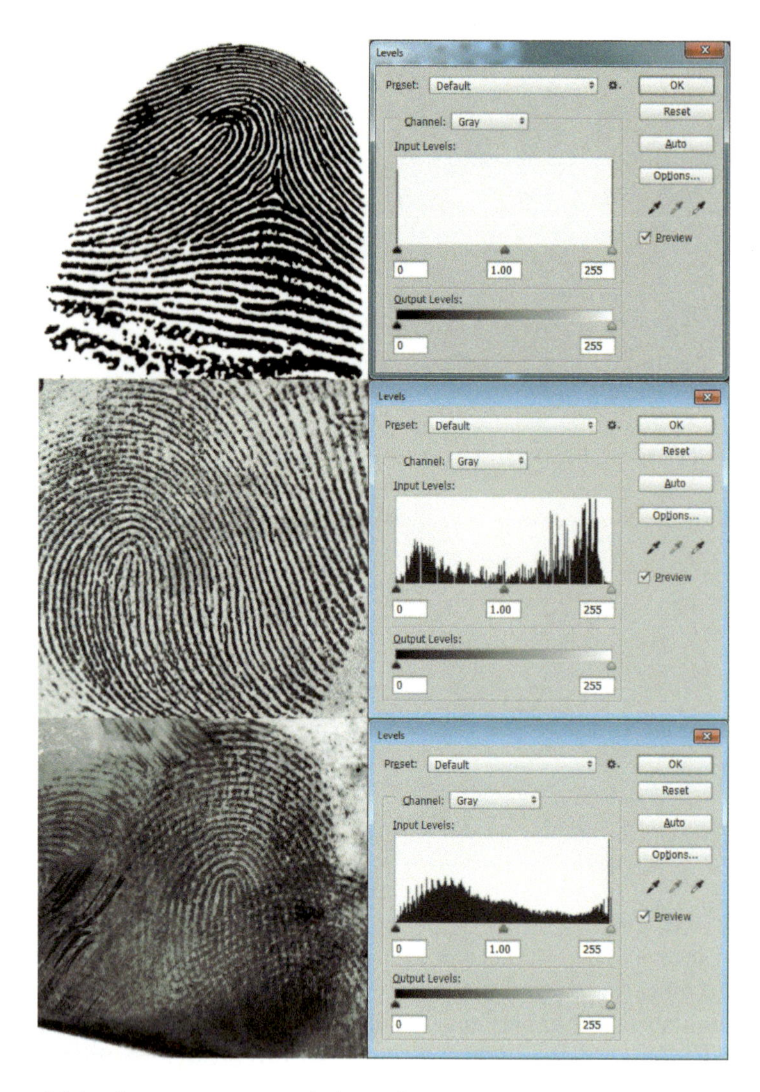

Figure 6.16 The comparison of three fingerprints with their corresponding Levels histogram beside them on the right. (Middle photo revised image courtesy of the *Journal of Forensic Science*, Issue: 2012-5-464, The Joy of LAB Color, Figure 1. All rights reserved. Bottom photo courtesy of York Regional Police. All rights reserved.)

plotted in the 0 column for black and another in the 255 column for white. That's interesting, but real-world fingerprints don't look like this. The second fingerprint is a processed version of the ninhydrin-developed print on a receipt we looked at in Chapter 3 (Figure 3.5). This image is from real casework, so it is not totally inconceivable that we may work with some images like this. Notice that there is a hill of pixels plotted at the dark end near black and another hill at the light end near white, but there are pixels of all tonal values plotted across the graph. The third image has plenty of tonal values of almost every range. To get a histogram reading of a specific area, make a selection, and open Levels to see a histogram of the selected area. If making any adjustments, remember to feather the selection to gently blend the effect.

Curves

The Curves dialogue box can be found by going to the Menu bar > Image > Adjust > Curves (keyboard shortcut: Ctrl—M).

The curves dialogue box is another graph mapping the tonal values of an image on a straight diagonal line. The steeper this line (called a curve), the more contrast between the tonal values *in the steep part of the curve*; the flatter the curve, the flatter the contrast in the tones of that area.

Click within the image, and a small circle appears on the curve where that tonal value is represented. You can set a point on this line by clicking on the curve directly or by Ctrl-clicking within the image to plot that value on the line of the curve.

Up to 14 control points can be added to the curve; this allows for very precise control when adjusting contrast. For example, anchoring the lightest ridge in an image means that while you adjust the furrow values to be lighter, this value cannot be lightened at the same time, thereby creating contrast between the ridges and furrows very specifically.

Edit Points

Use the "Edit points to modify the curve" option; tones can be plotted and then adjusted on the curve line. The other option "draw to modify the curve" is not an accurate or precise way of adjusting the contrast in an image for analysis.

Range of Tones

The vertical gradient scale (Figure 6.17/2) represents the output levels after changes have been made to image values, while the horizontal gradient scale represents the input or original image values before any changes were made. The input and output values displayed (before and after values) represent those of just the active point on the curve. Although you may plot up to 14 points, only one of them may be active at a time (shown as a solid black dot).

Eyedroppers

The eyedroppers work as they do in Levels. The black eyedropper can be clicked within an image to set a black point, just as the white eyedropper can

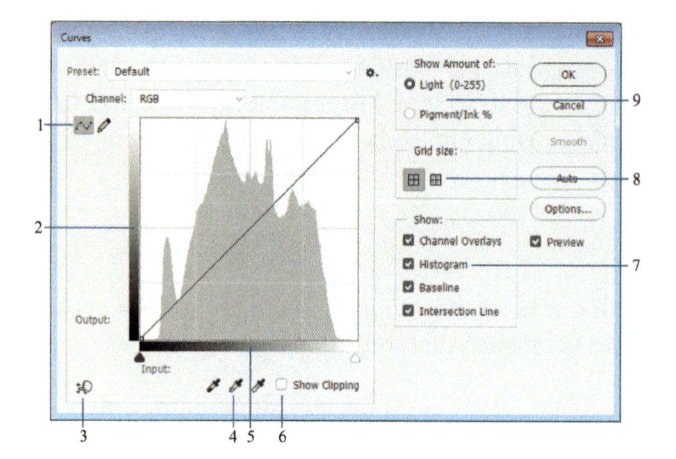

Figure 6.17 Curves dialogue box. (1) Edit points to modify the curve. (2) Range of image tones after changes. (3) On-image adjustment tool. (4) Eyedroppers. (5) Range of image tones prior to changes. (6) Show clipping. (7) Show channel overlays, histogram, baseline, intersection line. (8) Choose grid size. (9) Choose whether graph displays values of light or pigment.

be clicked within an image to set the white point. In a color image, the gray eyedropper can be clicked on a known neutral item to correct color casts.

Curves Options

The options in this area of the dialogue box simply refer to how you would like to customize it for your own ease of use. Display options default to Pigment/Ink when in grayscale or CMYK mode, reading the values of the image in percentages. If you prefer to read them as values of light from 0–255 (the RGB default), then change it from Pigment to Light. Either one works as well as the other; changing this display option merely flips the light side to dark and vice versa.

Show

Channel Overlays—Displays color channel curves superimposed over the composite curve.
Histogram—Shows the histogram in the background of the graph.
Baseline—Shows the original slope of the curve for reference.
Intersection Line—Displays the horizontal and vertical lines to align control points as you drag.

Grid size—Choose between a grid of lines in 25% increments or a detailed grid displaying gridlines in 10% increments.
Show Clipping—Check this box as you work in curves to preview areas of the image that are being clipped by adjustments.

On Image adjustment tool—Select to click and drag tones up or down directly on the image, changing the curve.

Making Adjustments in Curves

The next image we are going to work on, you have seen before: the ninhydrin-developed fingerprint on a receipt (Figure 3.5). The "a" channel has been inverted and applied to itself to give it a little contrast boost—a better starting point from which to begin our adjustments. Dragging the mouse over the image with the curves dialogue box open indicates where on the curve those tonal values lie.

Tip: *You can reference a real-time histogram as you make curves adjustments by opening the histogram palette. Go to Windows > Histogram (it will open on the right side of the desktop).*

There are three general methods that can be used to increase contrast in fingerprint impression images. Each method works fairly well as this image is roughly uniform in contrast and tone:

Curves Method 1—Pulling White and Black Points Straight Across

This image is extremely flat and is in dire need of some contrast (Figure 6.18). A simple way to increase contrast in an image using curves, especially one that does not have any shadows or highlights, is to drag the white point slider straight across the top toward the center and the black point slider across the bottom toward the center (see Figure 6.19). The tones of the impression still

Figure 6.18 The "a" channel in LAB of image 3.5—the "a" channel has been applied to itself in overlay mode and is ready for contrast adjustment. (Courtesy of the *Journal of Forensic Science*, Issue: 2012-5-464, The Joy of LAB Color, Figure 1. All rights reserved.)

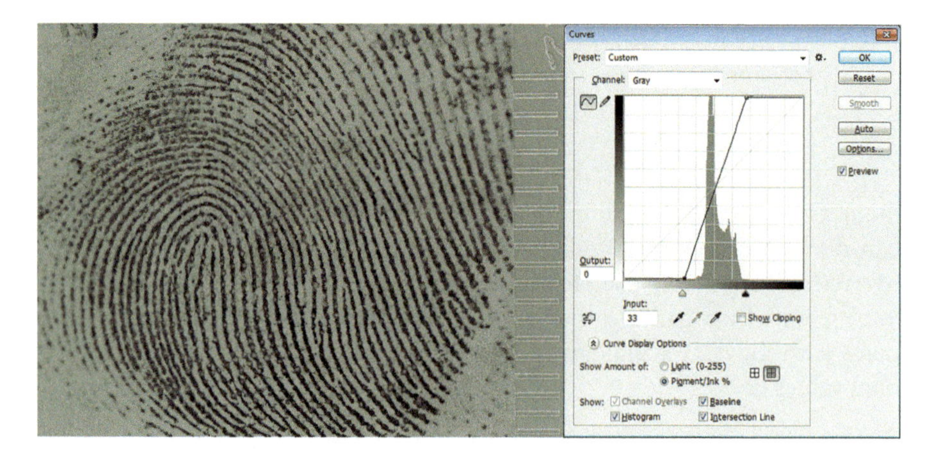

Figure 6.19 Contrast improved by dragging the white and black points in, to steepen curve. (Courtesy of the *Journal of Forensic Science*, Issue: 2012-5-464, The Joy of LAB Color, Figure 1.)

reside in the steep part of the curve, but now there is more contrast between the ridges and the furrows. Note that if you steepen the curve too much, clipping occurs.

Curves Method 2—The "S" Curve

Figure 6.20 shows a similar technique, but rather than harshly drawing in the white and black points, an S-curve is created. Find one of the very darkest

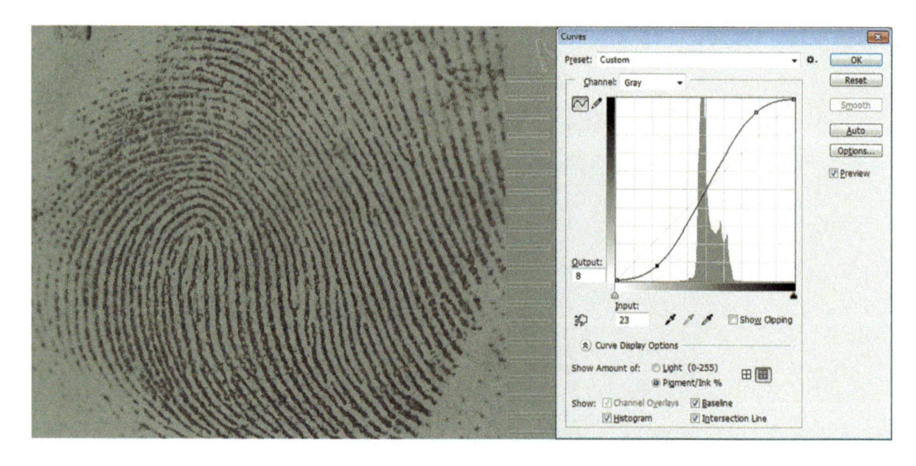

Figure 6.20 Contrast improved by creating the classic S curve shape. If any pixel values fall in the flattened part of the S curve at the top or bottom, they will decrease in contrast. (Courtesy of the *Journal of Forensic Science*, Issue: 2012-5-464, The Joy of LAB Color, Figure 1.)

points on the curve, plot a point just below it, and pull that point down a little. How much you pull that point down depends on the image; err on the side of *subtle*. Find one of the lightest points on the curve, plot a value point just above it, and pull that point up. What used to be a straight line now has a slight "S" curve to it, and the tonal values in the image are contained within the steep part of the curve, thereby increasing contrast. Any values that were darker or lighter than the points selected now reside within the *flat* part of the curve, exhibiting decreased contrast.

Curves Method 3—Precise Curves Adjustment

Figure 6.21 offers the most precise contrast adjustment possible. With Curves open, move your cursor over the image, and locate the *lightest or faintest* ridge tones. While holding the Ctrl key down, click the light ridge. That tonal value is plotted automatically on the curve with a black point (active). Find a good representation of the furrow near the light ridge, and Ctrl-click that point also to plot it on the curve. Activate the highest plot-point on the curve (click on it to activate it), and use your "up" arrow key on the keyboard to raise the point straight up. Next, activate your lower plot-point, and use the down arrow key to lower the point straight down. This method has targeted two specific tonal values and created further separation between them. Watch the light and dark areas of the image; many images may have light ridges that can be lighter than some furrows. Irregularities in the substrate, smudging of the fingerprint, and

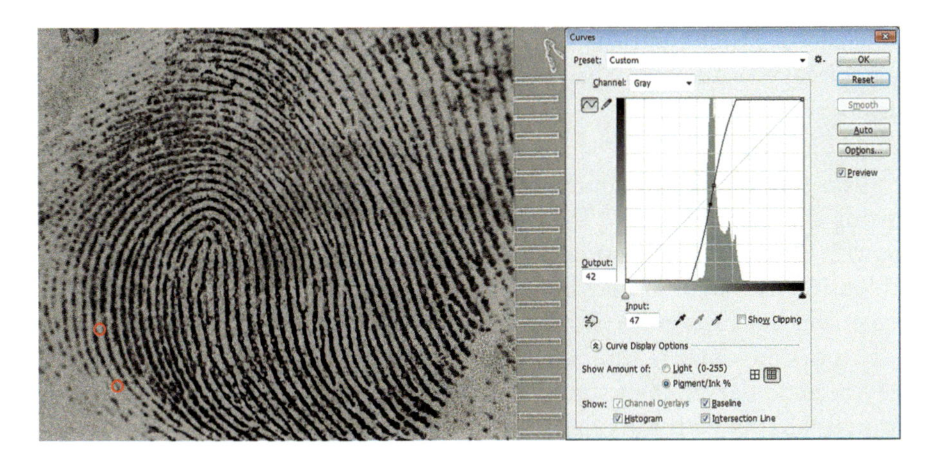

Figure 6.21 Contrast improved by Ctrl-clicking a light ridge (circled in, red), and then Ctrl-clicking a furrow (circled in red). Drag the top point straight up (or use the arrow buttons on the keyboard for accuracy), and drag the bottom point straight down. (Courtesy of the *Journal of Forensic Science*, Issue: 2012-5-464, The Joy of LAB Color, Figure 1. All rights reserved.)

other noise and dirt can complicate this process. If there are dark and light areas of a print, it might be a good idea to make a precise curves adjustment on feathered selections. If the fingerprint impression is very irregular in brightness, contrast, and ridge strength, the *Precise Curves* method might not be the right tool.

Whatever contrast adjustment method is chosen, use caution and watch all areas of contrast, while asking this question, "Has this adjustment made the impression better and clearer in all areas?"

Tip: *The Preview button (under the Cancel, Auto, and Options buttons) is available for all contrast adjustment tools. Click the button to toggle before and after before committing to an adjustment.*

The image can be brightened or darkened overall by placing just one point. Add a point in the middle area of the curve, and raise or lower it to lighten or darken, affecting midtones more than extreme darks and lights. Be aware that when the curve is modified with just one point, one side of the curve is steeper, thereby increasing the contrast of all the tonal values that reside there, while the other side is flatter, resulting in decreased contrast between the tonal values plotted in that part of the curve.

Shadows/Highlights

The Shadows/Highlights dialogue box can be found by going to the Menu bar > Image > Adjust > Shadows/Highlights. *It is not supported by adjustment layers.*

This tool is great for adjusting shadows and highlights separately, minimizing the risk of clipping the darkest and lightest portions of the image. As with all contrast adjustment tools, care must be taken; if wielded overzealously, the *midtones* may be in danger of being "clipped," appearing as posterized blocks of tone. Setting the *Amount* value too high can result in this type of crossover failure, when original light tones are darkened so much they become darker than some of the original dark tones in the shadows, for example. As with the other contrast adjustment tools, it is advisable to apply the Shadow/Highlight function conservatively to avoid loss of midtone detail.

Because any channel blending and pattern removal is done *before* the contrast adjustment stage, the following examples are all using 16-bit *grayscale* image (results are significantly improved when working with 16-bit images).

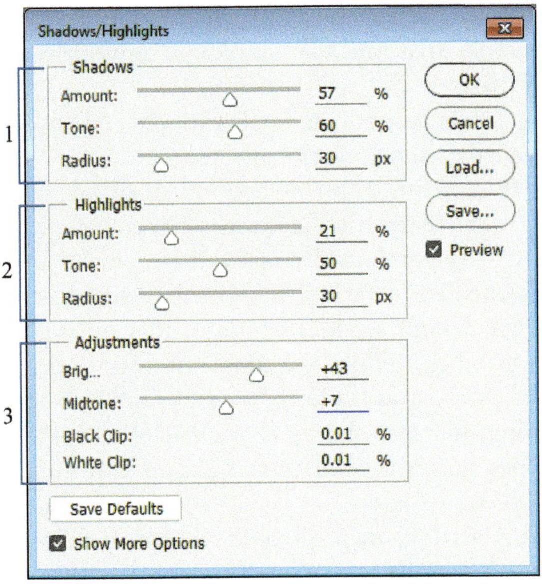

Figure 6.22 Shadows/Highlights dialogue box. (1) Adjust shadows. (2) Adjust highlights. (3) Adjustments.

The shadows and highlights dialogue box can be broken down into three main sections. If the dialogue box is only showing options for shadows and highlights, click the *Show more Options* box to expand your view. It opens with the shadow sliders set to lighten a silhouette, as the default settings assume an image is strongly backlit (Figure 6.22).

1. Shadows—Primarily affects the shadows and some midtones of the image; the highlights are not affected.
   Amount—This setting controls how much of a brightness adjustment is made to the shadow tones by percentage.
   Tone—The tonal width is the setting for defining the range of tones affected in the *shadows*, with the darkest being affected most. The default setting is 50%. If this setting brightens midtones too much, reduce the tonal width (try 25%) until only the dark areas of concern are affected (very dark blotchy areas of a fingerprint may respond better to a higher amount of brightness with a lower tonal width, preserving other shadow values). If the midtones require further brightening, increase the tonal width closer to 100%. Watch all area types closely as adjustments are made to ensure that detail is preserved.
   Radius—This setting determines the range of neighboring pixels affected. The brightness adjustment radiates out from the target

pixels as like a feather effect. A small radius may be likened to a small feather area, and a larger radius, conversely, feathers out to a larger area.

2. Highlights—Primarily affects the highlights and some midtones of the image; shadow tones are not affected.

   Amount—This setting controls how much of a brightness adjustment is made to the highlight tones by percentage.

   Tonal Width—The tonal width is the setting for defining the range of tones affected in the *highlights,* with the lightest being affected most. The default setting is 50%. If this setting darkens midtones too much, reduce the tonal width until only the lightest areas of concern are affected. If the midtones require further darkening, increase the tonal width closer to 100%. Watch all area types closely as adjustments are made to ensure that detail is preserved.

   Radius—This setting determines the range of neighboring pixels affected. The brightness adjustment radiates out from the target pixels as like a feather effect. A small radius may be likened to a small feather area, and a larger radius, conversely, feathers out to a larger area.

3. Adjustments

   Brightness—In grayscale mode, brightness is the first slider. Slide to the right to brighten the image overall or to the left to darken.

   Color Correction *(in color only)*—If the image is in RGB, CMYK, or LAB color modes, there is a color correction option rather than brightness, which increases or decreases saturation as you drag the slider.

   Midtone Contrast—As implied, this slider adjusts contrast to the midtone values within the image.

   Black Clip and White Clip—The black and white clip values determine how much of the shadows and highlights are clipped to the new black and white level in the image. The remaining tonal values are redistributed between the brightness values of 0–255. If the value is too high for either the black or white clip, detail is lost in the shadows and highlights. Keep these settings at the default values of. 01% to avoid clipping dark or light areas in the image as you adjust.

   The dark, blotchy fingerprint image of Figure 6.23 is greatly improved using shadows/highlights (see settings of Figure 6.22). 60% of the shadows (tonal width) were brightened 57% (amount). Radius—30-pixel feather. Then 50% of the highlights (tonal width) were darkened 21% (amount). Radius—30-pixel feather. Brightness was increased to 43, and midtone contrast bumped up to +7.

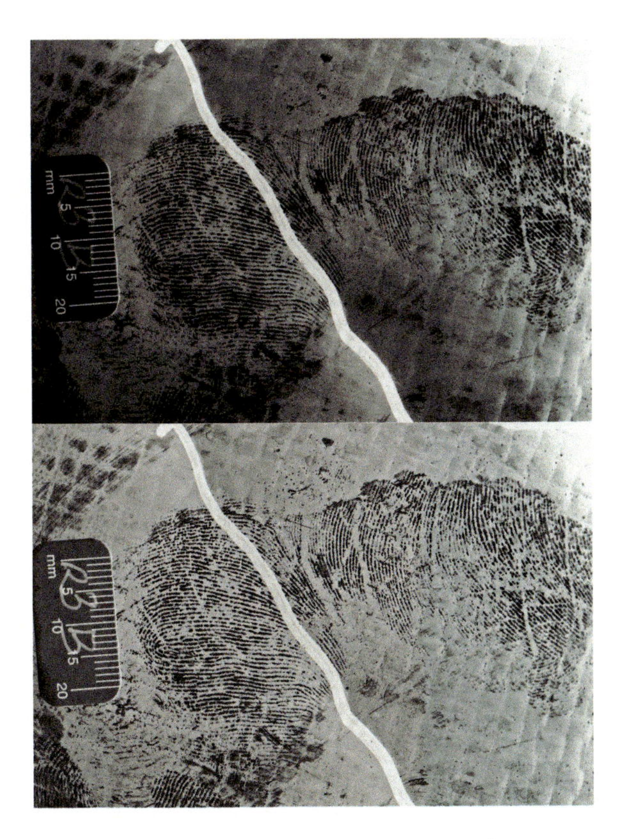

Figure 6.23 Top: Fingerprint after channel blending and converting to grayscale in need of contrast adjustment. Bottom: Fingerprint after applying shadows/highlights. (Courtesy of Vince Eagan, York Regional Police. All rights reserved.)

Sharpening Techniques

While sharpening functions are effective at increasing the contrast detail of ridges and furrows in a digital image, it may also be said that the sharpening tools in Photoshop are a double-edged sword. The processing tools of any software solutions have the potential to create visual havoc with an image, making details difficult to discern, and sharpening tools are no exception. How can these hazards be avoided? The following sharpening techniques address how to choose settings to avoid overprocessing.

How Does Sharpening Work?

The unsharp mask algorithm looks for edges within an image. Within the dark side of an edge, a dark halo is created, and within the light side of the edge, a light halo is created, boosting the contrast of the detail. The size and

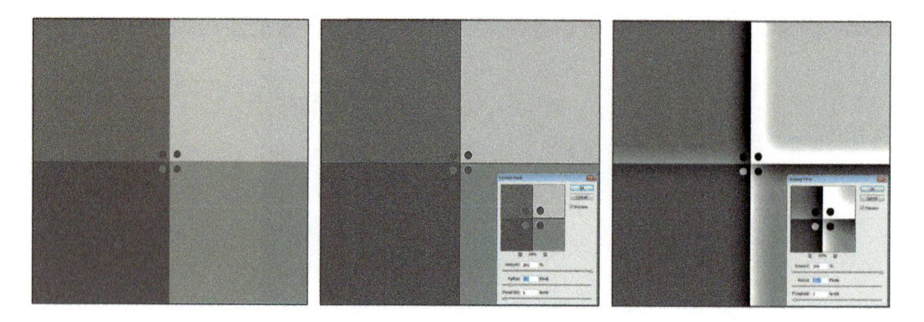

Figure 6.24 The effects of sharpening with both a high amount, low radius, and low amount, high radius.

strength of the halo effects can be tailored specifically for the image being sharpened. There are two general approaches to the sharpen/contrast techniques being applied: High amount/low radius, and low amount/high radius. The amount value refers to the strength of the effect, and the radius refers to the range of pixels on either side of the edge that are affected (how wide the halo is).

Figure 6.24 shows the effects of both techniques. The square on the far left is the original set of squares created for this illustration, each with a different tonal value inside. Close to the center, there are circles filled with another tonal value to illustrate the sharpening effects of smaller details in an image.

Center image (high amount—500%, low radius—1.8 px)—The light and dark halos can be easily seen to create contrast between the edges. In this example, unsharp mask has done a great job accentuating edges of both large and small detail; however, be aware that this technique has the potential to grossly sharpen small noise in the image, such as dust, dirt, metallic paint flecks, or other minute artifacts. This is probably not desirable, in which case, you may prefer to set the amount lower and increase the radius.

Left image (low amount—500%, high radius technique—36.7 px)—Although it has been described as a "low amount" technique, for illustrative purposes, the Amount is set at 500% so that the halo effects may be plainly observed here. The radius is very high, with obvious halo banding along the edges. Note the effect on the inner circles, as some of them are starting to blend into the halo banding. This technique helps to find the ideal radius for the image detail and create contrast between the ridges and the furrows without losing the detail of pores or incipient ridges. The *amount* at 500% aids in visualizing the contrast effect as the radius slider is adjusted. Once the desirable radius is set so that ridges are clearly defined without loss of detail, reduce the *amount* to something more realistic. Again, find the small details within the ridge, and watch those very closely as you adjust the radius. If the small detail starts to disappear in the halos, bring the radius back down to a lower value until detail is preserved.

Understand that the purpose of using the unsharp mask function is not to actually sharpen the image; a blurry, out-of-focus image cannot be made "in-focus" again. Forensic imaging specialists who understand this tool may use it for the contrast adjustment of edges on an image that is already in focus.

Unsharp Mask

The Unsharp Mask dialogue box can be found by going to the Menu bar > Filter > Sharpen > Unsharp Mask. It is not compatible with adjustment layers (Figure 6.25).

Unsharp Mask features the following input settings:

Amount—The higher the percentage, the stronger the effect.
Radius—The range of pixels on either side of the determined edge that is affected.
Threshold—The level of difference in tonal values between pixels before they are considered to be edge pixels. Keep this value low, roughly between 0 and 5.

As previously mentioned, there are two general methods of sharpening with this tool:

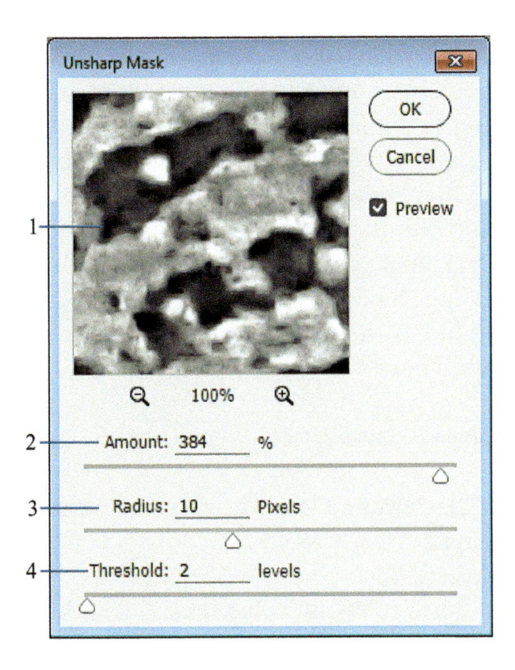

Figure 6.25 Unsharp Mask dialogue box. (1) Preview window. (2) Amount. (3) Radius. (4) Threshold.

Method 1: High amount (around 500%), low radius (1-2 pixels); a good technique for sharpening the tiny detail within an image (such as pores). If there is any noise present, this setting may increase chatter to the detriment of image clarity. Keep threshold between 0-5 levels of difference for edge finding.

Method 2: Lower amount (30%–150%), high radius (5–30 pixels)—To find just the right radius, set the amount to 500% temporarily. Slowly move the radius slider up until clear definition of the ridges are visualized without loss of finer detail. Reduce the amount to a desirable contrast strength.

Reduce Glowing Halo

While unsharp mask does offer a good deal of control over the strength and size of these effects, the white halos tend to be distracting to the eye, much more distracting than the dark halos. The first technique for toning down the "white halo effect" renders those bright halos down to a tolerable degree without affecting the dark halos. Open Image 6.26 to follow along.

- Create a duplicate layer of the image you wish to sharpen (Ctrl J for PC, Command J for Mac).
- Select the top layer. Go to Menu > Filter > Sharpen > Unsharp Mask.
- Choose your Amount, Radius, and Threshold (0–5). Click OK.

The Unsharp Mask filter has been applied to the top layer only, and the bottom layer remains unaffected. The intent at this point is to apply the bottom layer to the top layer in darken mode. Darken mode instructs Photoshop to *"compare the pixel values between the layers and use the darker of the two."* This keeps all the dark halos and dark contrast effect inside the ridges and eliminates the light halos in the furrows completely until we choose a percentage. Setting the percentage between 50 and 75% keeps some of the light halo effect but to a diminished degree, which is much more acceptable to the human eye, making the flow of the ridges easier to follow. Here's how:

- With your top layer still selected, go to Menu > Image > Apply Image.
- Layer—Select "Background" from layer menu to choose the unsharpened bottom layer.
- Blending—Darken.
- Opacity: Defaults to 100%, meaning that all your highlight sharpening halos are eliminated. If you wish to retain your highlight sharpening halos at a reduced strength, try 50–75% opacity.
- Click OK (Figure 6.26).

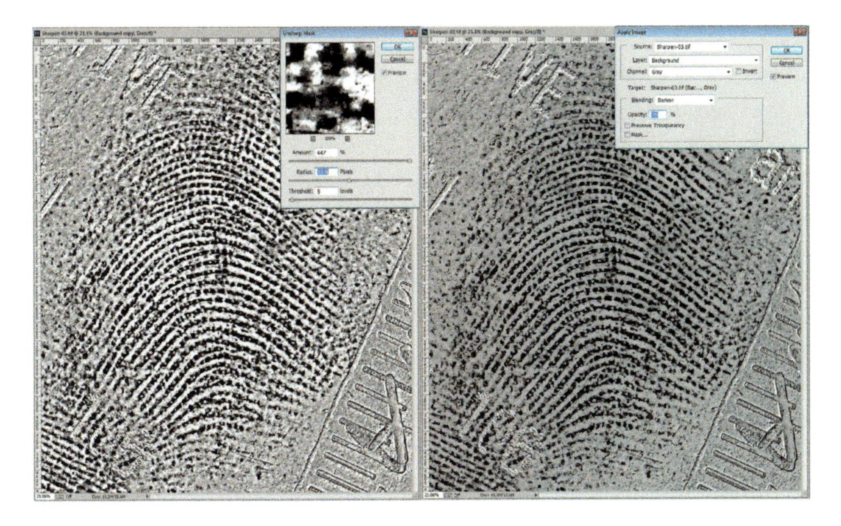

Figure 6.26 On the left, step 3 (Unsharp Mask) is illustrated with exaggerated settings to show the effect clearly. On the right, step 4 shows the apply image in darken mode settings and its result. (Courtesy of Andrew Cholmondeley, York Regional Police. All rights reserved.)

Figure 6.27 Layers palette.

This method offers total control of halo-highlights in the final image. Further, the sharpened version is on the top layer, with the original on the bottom. The opacity of the sharpened layer may be reduced if desired (Figure 6.27).

Smart Sharpen Filter

The Smart Sharpen Filter can be found by going to the Menu bar > Filter > Sharpen > Smart Sharpen Filter. *It is not compatible with adjustment layers.*

The Smart Sharpen Filter in Photoshop has settings within it to control the dark and light halos in a way that is similar to the technique we just applied. While the results are similar, they are not the same. I find I prefer the flexibility of the unsharp mask combined with the apply image technique. Smart sharpen does a comparable job, if a little subtler.

Amount—The higher the percentage you choose here, the stronger the effect.

Radius—The range of pixels on either side of the determined edge that are affected.

Reduce Noise—Reduce unwanted noise without affecting edges.

Remove—Choose a sharpening mode between Gaussian blur, lens blur, and motion blur. Gaussian blur sharpens identically to the unsharp mask filter. Lens blur allows you to sharpen typical optical lens blurring. It provides finer sharpening of detail and does a better job of reducing sharpening halos than the Gaussian blur setting. Motion blur works best when you want to remove small amounts of motion blur that resulted from camera movement. Click the wheel to the right of the angle field to set the direction of the motion that needs to be removed (not applicable in forensic photography of impression evidence because of the use copy stands and tripods).

If this is all your smart sharpen panel offers in the way of settings, then the panel must be expanded to show more options; click the > Shadows/Highlights triangle underneath the *Remove* setting.

CS6 has a *More Accurate* option: It processes the image more slowly for a more accurate removal of blur.

Shadows

Fade Amount—0% does not fade the shadow halo at all. 100% fades the shadow halo completely.

Tonal Width—A lower percentage value limits sharpening to darker tones in image. A higher value expands sharpening to include a wider range of dark tones.

Radius—The range of pixels on either side of the edge that is affected.

Highlights

Fade Amount—0% does not fade the highlight halo at all. 100% fades the highlight halo completely.

Tonal Width—A lower percentage value limits sharpening to the lightest tones in image. A higher value expands sharpening to include a wider range of brightness tones.

Radius—The range of pixels on either side of the edge that is affected.

For Figure 6.28, the settings in Smart Sharpen are the same for the amount and radius as were used for example of Figure 6.26 in an effort to make the effects as similar as possible for comparison purposes. Reduce noise is set to 50%, shadows are left at the default of 0% fade amount to leave them as is, and the fade amount of the highlights is set to 75% (tonal width 100%). Although

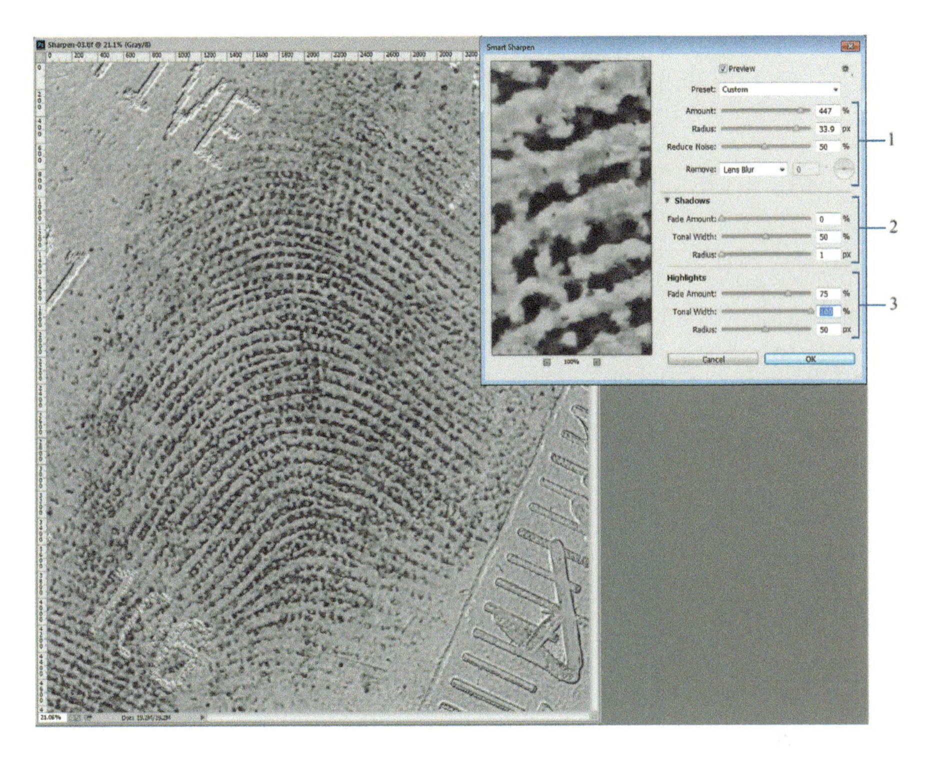

Figure 6.28 Smart Sharpen filter and the subtle results on the fingerprint in Figure 6.26. (Courtesy of Andrew Cholmondeley, York Regional Police.)

some effort was made to recreate the same effect as with the unsharp mask technique, this result is somewhat more subdued. Both are improvements on the fingerprint detail of the image. In the end, how much or how little to sharpen is subjective, provided the shape and contour of ridges, pores, and incipient ridges are maintained. My advice is to use a gentle hand and not overdo it.

High-Pass Sharpen with Adjustment Layers

While the first two sharpening techniques are *not* compatible with the *Adjustment Layers* technique of recording processing steps, there is another sharpening technique that is. A high-pass filter may be used on a duplicate layer to create a sharpening and contrast adjustment to Figure 6.26. The high-pass filter seeks edge details within the pixel radius specified and suppresses the rest of the image. Edges are defined with primarily darker tones (and some lighter tones) in its own layer, while the rest of the layer remains 50% gray. When blended with the image in overlay mode, values that are 50% gray in the high-pass layer affect no change. Darker and lighter definitions in the high-pass layer are blended with the layer below. This creates a subtle sharpening effect that improves the contrast of ridge edges in the image.

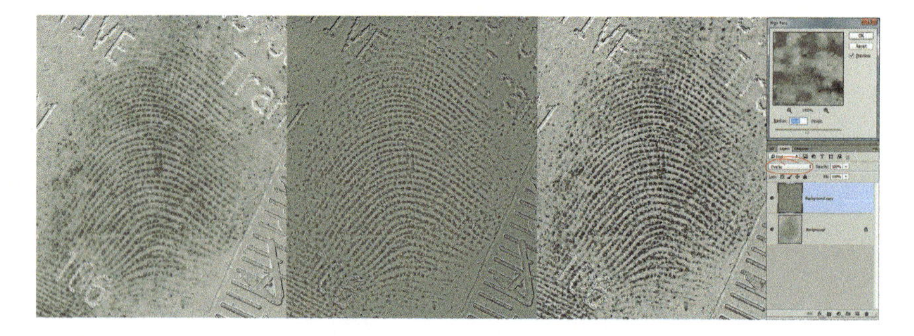

Figure 6.29 Left: Original image. Middle: High-pass effect. Right: Image after applying the high-pass filter to the top layer and changing the blending mode to overlay. Side boxes: The high-pass dialogue box and the layers palette after setting the mode to overlay. (Courtesy of Andrew Cholmondeley, York Regional Police. All rights reserved.)

If recording processing steps using Adjustment Layers, a composite layer should be created on top of all image layers. Since sharpening is typically the last step in image optimization, there are likely other adjustment layers above the original background image. Ensure the top layer is active (highlighted), and hold down the keyboard keys Shift, Ctrl, Alt, and then press "e." A new composite layer will appear on top of all layers in the file. It is essentially a flattened version of the layers below it, while maintaining the adjustment layers below for reference.

- Duplicate the image layer or composite layer just created (Ctrl—J for PC, Command—J for Mac). Confirm that the top two layers (if using adjustment layers, there may be many other layers) are identical composite layers.
- Select the top composite layer. Go to Filter > Other > High Pass.
- Set the pixel radius anywhere from 10-pixel radius for a subtle sharpen effect, to 40 pixels or more for a more pronounced effect. The higher the resolution of the image, the higher the radius that is required. Make a note of the radius. Click OK.
- Above the top layer in the layers palette, change the layer blend mode from normal to overlay.
- Re-name the high-pass layer by double clicking the layer name. Name it intelligently to assist your understanding of the work done at a later date. (e.g., high pass—radius 34.6) (Figure 6.29).

Advanced Selections

This is an image selection technique that has been saved until this chapter, in order to have introduced channels, selections, and contrast adjustment tools

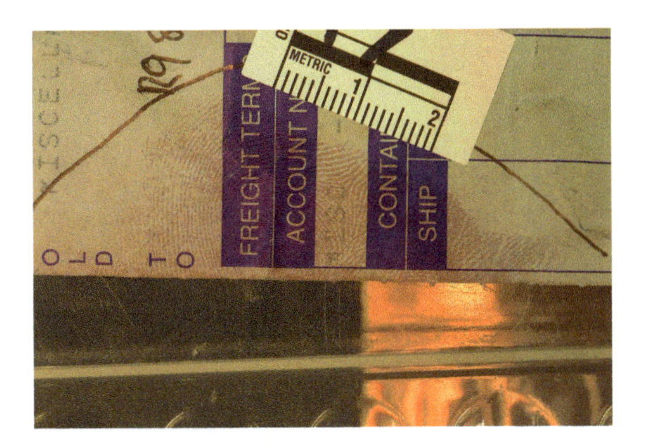

Figure 6.30 Ninhydrin-developed image as captured with the camera. (Courtesy of York Regional Police. All rights reserved.)

first. The following is a technique that is not required very often, but nonetheless, it is a good one to know (Figure 6.30).

This image has two very distinctly different color areas: the paper background of the shipping label and the blueprint portion, where there is more noise, and the ninhydrin print has developed a little differently than the more porous portion of the paper. Scrolling through the RGB color channels, it is obvious that while the red channel contains a modest amount of fingerprint signal, the green channel is better, and the blue channel is even better than that. However, even in the blue channel, it would be desirable to treat those two color areas separately, and that requires making a selection. Is the polygonal lasso tool appropriate? It's not impossible, but there is a better way.

What Is a Selection Mask?

Upon reviewing the color channels, it becomes apparent that the red channel holds some potential for creating a selection mask. This means that the red channel may be modified to create a pure black and white representation of the color areas in our image. Once that is done, adjustments may be made on each of those areas separately, with ease. Why the red channel? The red channel contains strong information for each of those color areas, with minimal fingerprint detail.

Make a selection mask using the red channel following these steps:

- In the channel's palette, select the red channel.
- Open the Levels dialogue box (Image > Adjustments > Levels).
- Adjust the white and black point sliders to make the area representing the blue label pure black and the rest of the image pure white.
- At the bottom of the Layer's palette, click the "Load channel as selection" button (see Figure 6.31).

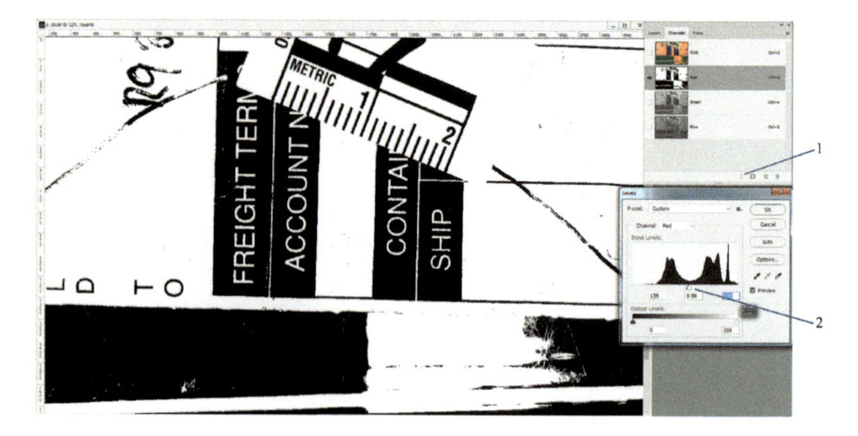

Figure 6.31 The red channel is activated, and the Levels dialogue box is open. (1) "Load channel as a selection" button. (2) The black and white points have been dragged in on the Levels histogram to make the red channel pure black and white (black point—136, white point—138).

The white part of the *mask* just created is now selected; there should be *marching ants* around the perimeter of the selection. Now activate (click) the blue channel. Contrast adjustments may now be made to the paper area of the image. To work on the blue-inked portion of the image, go to the Select drop-down list on the menu bar, and choose *Inverse* (Menu: Select > Inverse). When contrast adjustments for both areas are completed, convert to grayscale and save as an optimized or enhanced copy of the original file (Figure 6.32).

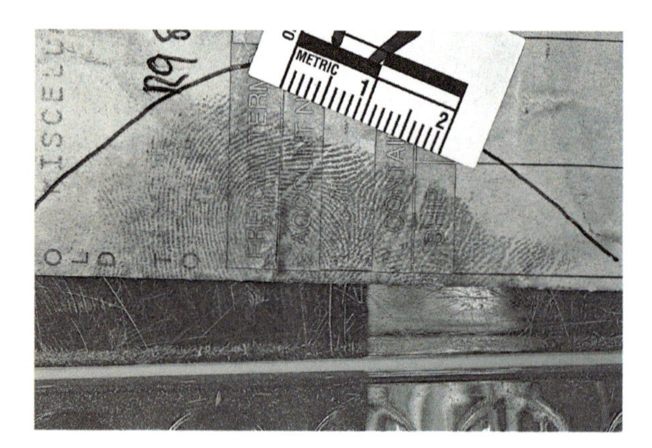

Figure 6.32 The result: Both areas have been adjusted for contrast separately using apply image (applying the blue to the blue in overlay mode), noise reduction, shadow/highlight, levels, and unsharp mask. Only the blue channel was used to process the fingerprint signal. (Enhanced version of Figure 6.30 courtesy of York Regional Police. All rights reserved.)

Review Questions

1. Most selections require this to ensure a soft blending of the effect in the image.
2. True or False: All contrast adjustment tools have the ability to clip detail, especially from the shadows and highlights.
3. If you are working with a grayscale image, how do you adjust contrast with levels while ensuring that there are no highlights or shadows clipped from the area of interest?
4. How many points can be adjusted on a curves graph at a time?
5. How can the distracting highlight halos be minimized after sharpening an image for analysis?

The Approach
Developing Enhancement Strategies for Images Intended for Analysis

7

Experience is simply the name we give our mistakes.

Oscar Wilde

Review: Putting It All Together

There have been many techniques covered in this book, and at this point in our workshop, someone often asks, "There are so many options and tools—how will I know which ones to apply?" It is a good question; it doesn't matter how many tools you have in your toolbox if you don't know when, why, or how to use them. The answer: practice, experience, and knowing that there are likely several different techniques that may be employed to come up with comparably similar results. So, take the pressure off yourself, and realize that finding a processing technique that works isn't like finding a needle in a haystack, but rather it is more like choosing one of several available methods. As you work through possible techniques, look at the detail in the image and ask yourself, *"Is it better? Is it clearer? Is there any detail lost/clipped from the shadows and highlights? Can I explain what I've done?"*

The purpose of this chapter is to summarize the four general enhancement steps necessary to approach any image enhancement, followed by a series of practical exercises that review what we have learned in previous chapters.

The Four General Enhancement Steps

Each image must first be evaluated and a unique processing plan developed. The objective may be broken down into four general steps: adjust RAW image in Camera Raw (or other processing software) to best visualize detail in shadow, highlight, and midtone areas of the impression, prepare and size for AFIS (1:1), minimize distracting background patterns (image subtraction, FFT, or channel blending), and last but not least, adjust for brightness and contrast.

This simplifies the work of processing, as each objective may be addressed in order, one at a time:

1. *Adjusting the RAW image:* It is recommended to set the workflow options within the Camera Raw dialogue box to open images in 16-bit format for images of very high contrast, such as those photographed with a forensic light source, as detail may be made more visible by lightening the shadows and/or darkening the highlights (review Chapter 2: RAW File Formats and Image Processing).
2. *Prepare for AFIS:* Calibration 1:1 is a step-by-step task, outlined at the end of Chapter 2 (review Chapter 2: Steps to Calibrate Image 1:1).
3. *Minimizing background noise:* This step may or may not be necessary—it depends if there actually *is* distracting noise or patterns in the substrate. First, determine if there are any distracting background or noise elements that are detracting from the clarity of the signal detail. The following are examples of possible image enhancement scenarios and the strategies to proceed at this step:
   a. No distracting background elements/noise: Proceed to contrast adjustment step.
   b. Neutral/gray tones for signal *and* noise: If the entire image is neutral in tone, such as a fingerprint dusted with gray powder on a black binder (pebbly texture) or black powder dusted on a gray metallic car, for example, there may be very little that can be done. Even the smallest amount of contrast adjustment on this type of image can make the noise worse. Here are some options for this type of impression evidence:
      i. *Proactive:* Photographer dusts the impression with colored powder to differentiate it from the neutral tones of the substrate.
      ii. *Proactive:* Photographer shoots the image to the best of their ability using a *tripod*, cleans off the substrate, and photographs the background for image subtraction later during the computer processing stage. *Note:* Dirt and powder residue existing on the substrate will also be part of the final result (review Chapter 4: Erasure Subtraction Technique).
      iii. *Proactive:* Photographer stains the impression and photographs with a forensic light source using filters. This increases the possibility of clarifying a great deal of detail (if it's there).
      iv. *Reactive:* Fast Fourier transform may be used to remove repetitive background patterns in *some* cases (review Chapter 6: Fast Fourier Transform).

   You'll notice that the successful outcome for processing an impression comprised of neutral tones deposited on a noisy substrate, also comprised of neutral tones, often depends on the photographer making a thoughtful choice *before* taking the shot (proactive). Otherwise, there just may not be a solution to clarify that impression detail. Training forensic photographers to think through possible scenarios to get the best possible image capture in the first place is crucial to the success of visualizing image detail later.

c. *Colorful signal/colorful substrate and noise:* There is no guarantee that the color in the image is going to be able to assist with enhancing the impression, but often enough it does. Review the color channels in RGB, CMYK, and LAB:
    i. *Proactive:* The photographer chooses colored powder for dusting a latent impression that is complimentary to the substrate so that it can be differentiated more easily at the computer processing stage.
    ii. *Proactive:* Again, the photographer shoots the image to the best of their ability using a tripod, cleans off the substrate, and photographs the background for subtraction during the computer processing stage.
    iii. *Proactive:* Photographer stains the impression and photographs with a forensic light source using filters.
    iv. *Reactive:* A channel is found to exist in the image that significantly improves the clarity of the impression while muting the background noise—select the channel and convert to grayscale. Proceed to contrast adjustment.
    v. *Reactive:* A channel exists that improves the clarity of the impression but still contains noise—if another channel exists that contains primarily the noise, go to Calculations, and attempt to subtract the noise channel (Source 1) from the channel containing the impression (Source 2). This technique takes practice. Experiment (review Chapter 4: Channel Blending in Adobe Photoshop).
    vi. *Reactive:* Fast Fourier transform may be used to remove repetitive background patterns in some cases (review Chapter 5: Fast Fourier Transform).

4. *Adjusting contrast:* Once any possible channel blending, image subtraction, or FFT (pattern removal) techniques have been completed to optimize the signal-to-noise ratio, then the contrast may be adjusted. Caution must be taken to not clip the detail in the highlights and shadows. The expectation should *not* be to make the impression look like a rolled print—that is seldom possible. The tools that may be used to adjust contrast and brightness are: Levels, curves, shadows/highlights, and sharpening; remember that

it is not necessary to use each and every contrast adjustment tool for every image but to choose the tool believed best suited to the task. Again, practice dictates what works best (review Chapter 6: Contrast Adjustments).

Review Exercises

These exercises are intended for practicing the techniques covered in the previous chapters of this book. The tools and techniques used in this chapter are be demonstrated in great detail, but if you are struggling, you may wish to revisit those chapters to refresh your memory as you work.

Chapter 7—Exercise A (Image 7.1)

Figure 7.1 can be found on your disc as both the JPG and the RAW file. This is an exercise in preparing your RAW image with Camera Raw to maximize the visibility of detail in all areas of brightness. Sometimes, especially when photographing with a forensic light source, increasing the visibility of ridge detail in an image means decreasing the contrast (brightening the shadows and darkening the highlights). Open Figure 7.1, and adjust exposure, contrast, highlights, and shadows. Find that sweet spot where you can best visualize the

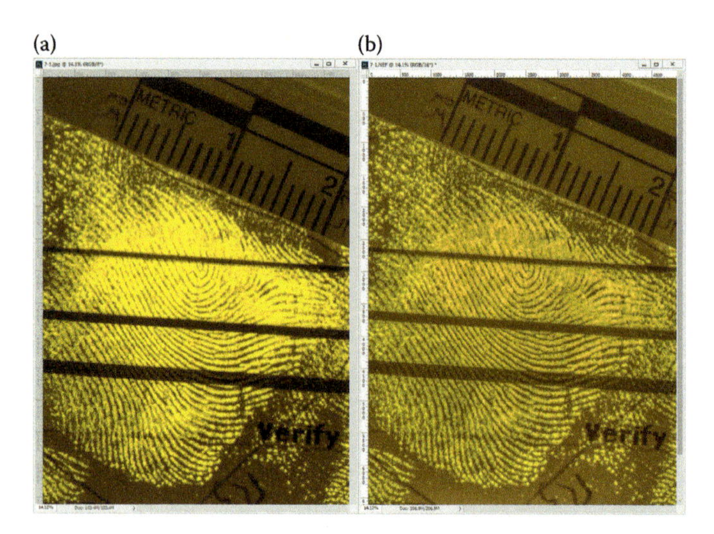

Figure 7.1 Starting with a RAW image: (a) Auto settings in the Camera Raw dialogue box. (b) Using the sliders in Camera Raw, manually lower exposure, lower contrast, lower the highlight values, and increase the shadow values. (Courtesy of York Regional Police. All rights reserved.)

detail of the fingerprint. Watch for ridge detail in the brightest areas and in the darkest shadows; the goal is not to create an image with high contrast, but rather to be able to visualize the ridges that are present but hard to see.

Chapter 7—Exercise B (Image 7.2)

Figure 7.2 is a good example of an image that lends itself well to many different enhancement strategies. Open it up and try everything!

Now that you have worked with this image a bit, here are several examples of possible approaches that seem to work well with this image:

Strategy Number 1

Green channel (or the magenta channel in CMYK color mode) alone.

If selecting the green channel alone is what you feel comfortable with, then do it. It is a definite improvement from the original image. You must be prepared to explain what you did and why you did it in court, so work within your area of understanding and comfort.

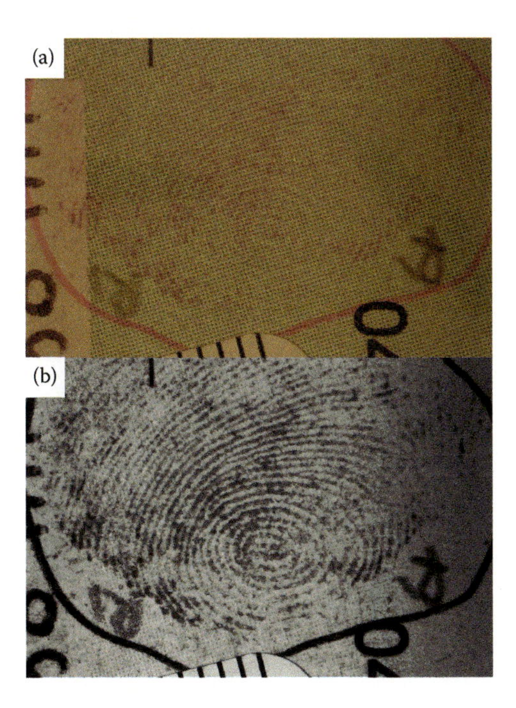

Figure 7.2 (a) A ninhydrin-developed print on green-dotted printed paper. (Courtesy of the *Journal of Forensic Science*, Issue: 2012-5-464, The Joy of LAB Color, Figure 7.1. All rights reserved.) (b) Results after blending the "a" channel of LAB with the green channel in Calculations. Note that the dotted pattern is still visible, as we have not succeeded in eliminating it altogether.

Strategy Number 2

Using Calculations (Image > Calculations): Red channel (Source 1), subtracted from the green channel (Source 2), opacity around 30–35, offset around 50–70. Remember that if you are subtracting, it is important to adjust your offset to visualize your results!

Strategy Number 3

Duplicating the image (Image > Duplicate), convert one of the versions to LAB color mode. Apply the "a" channel to itself (Image > Apply Image) in overlay mode. Open Calculations (Image > Calculations), select the green channel (Source 1) and the "a" channel (Source 2), and set the blending option to overlay. Invert the "a" channel (check the invert button beside the "a" channel option in Source 2). Why? Because the "a" channel is sporting light ridges/dark dots, and the green channel has dark ridges/dark dots. Our goal is to achieve dark ridges and minimize the dots. Once the "a" channel is inverted, it now has dark ridges/light dots. This technique blends dark ridges together, and dots will potentially cancel each other out, since they are light in one channel and dark in the other.

Strategy Number 4

Convert the image to LAB color mode. Apply the "a" channel with itself in overlay mode. In Calculations, either multiply or overlay the lightness channel with the "a" channel (invert button checked).

Strategy Number 5

Select either the green channel in RGB, or convert to CMYK, and select the magenta channel. Convert to grayscale. In Image-Pro Premier, use fast Fourier transform to remove unwanted repetitive pattern from the background.

The above examples are only five possible strategies that may be used to get started on this image; there are many more that would work just as effectively. After blending channels to optimize the fingerprint to the best of your ability, move on to the contrast adjustment step (levels, curves, shadows/highlights, or unsharp mask).

Chapter 7—Exercise C (Image 7.3)

Open Figure 7.3 in Camera Raw. Ensure the workflow options are set to open image as a 16-bit file. If not, you may recall from Chapter 2 that you may set the Camera Raw dialogue box to open all RAW images as 16-bit files by clicking the file information link at the bottom of the dialogue box to open

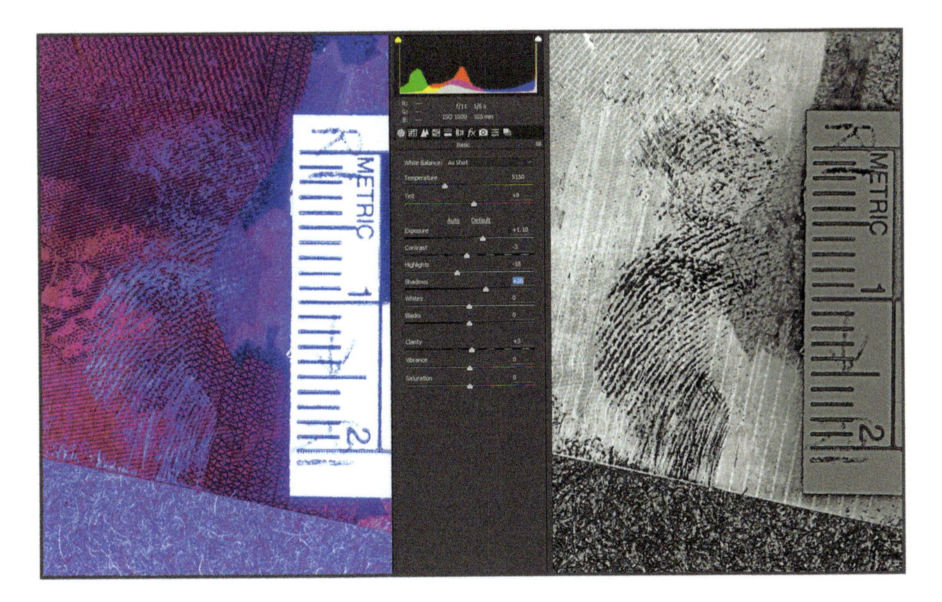

Figure 7.3 (Left) Color image resulting from Camera Raw settings (center) that tone down the highlights and increase the exposure in the shadows to a degree. (Right) Grayscale image after enhancement (red channel noise subtracted from blue channel fingerprint signal). (Courtesy of Jason Goodfellow, York Regional Police.)

"Workflow Options" (this link may be recognized at the middle/bottom of the box, and it underlines a list of information: color space of the file, bit-depth, pixel dimensions, file size, and how many pixels per inch). Once the bit-depth is set to 16 bits in workflow options, it remains set that way until it is manually changed.

Figure 7.3 depicts the Camera Raw settings used when opening the image. Image on the right: The noise of the red channel was subtracted from the blue channel (both channels had to be inverted to achieve black ridges), opacity set at 45%, and contrast adjustments applied using shadow/highlight and curves. Again, that is not to say that there are not many other ways to process this image. For example, in LAB color mode, the "b" channel alone looks promising. There are often many paths that can be taken to improve the signal to noise ratio.

Chapter 7—Exercise D (Image 7.4)

Figure 7.4 is brown paper treated with ninhydrin. The texture and fibers of the bag make the ridge detail very hard to see clearly; it doesn't help that the color values are very close between the brown paper bag and the ninhydrin-developed print.

Figure 7.4 An image on rough brown paper treated with ninhydrin. (Courtesy of Mark Hoekstra, York Regional Police.)

Upon reviewing the RGB channels, it becomes apparent that the ridge detail, weak though it may be, is most prevalent in the green channel, while the texture and fibers of the substrate are most prevalent in the blue channel.

Selecting the green channel by itself may offer some nominal improvement to the clarity of the print, but if we delve into the Calculations dialogue box and attempt to subtract the blue channel (Source 1) from the green channel (Source 2), the results show a faint but much clearer picture of the ridge detail we seek. This is not a perfect image subtraction because we are working with channels that all carry some of the contrast and detail of the composite image as a whole, so the result still shares some of the texture, and making contrast adjustments exacerbate the distracting substrate texture further, so tread softly there. It is more important to see the ridge flow as clearly as possible even with a softer image of lower contrast than it is to increase contrast to the degree that the image as a whole looks like oatmeal, with subtle detail either clipped or obscured with noise.

The author did try converting to LAB to see if there was some advantage to the color channels there, and while the "b" channel held some possibilities, it just didn't give enough detail and clarity to the print, so that option was discarded.

Chapter 7—Exercise E (Image 7.5)

Sometimes the lifts are not as successful at depicting the ridge detail of a print as the photograph, but the substrate in the photo is unfortunately

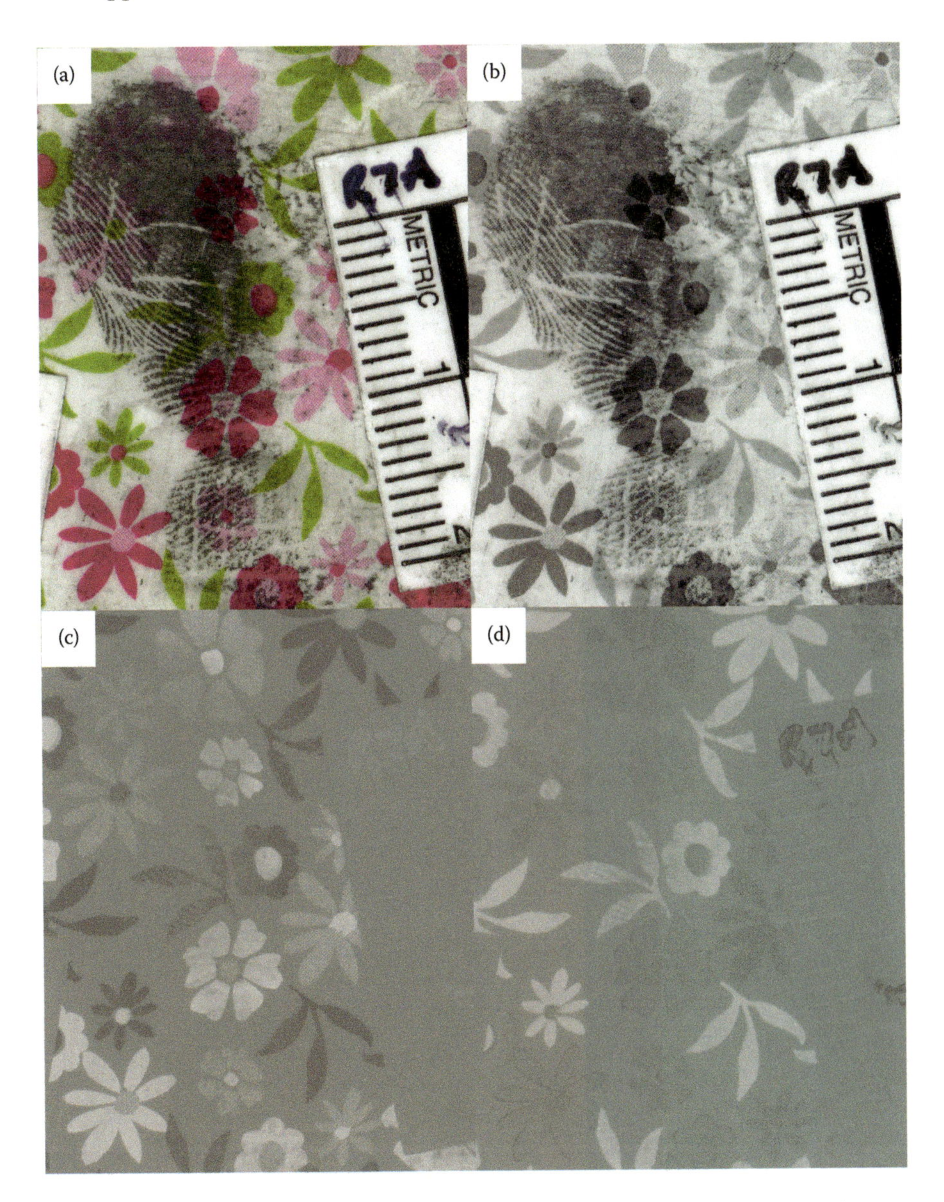

Figure 7.5 (a) An image dusted on a colorful and flowery substrate. (b) Lightness channel. (c) "a" channel. (d) "b" channel. (Courtesy of Andrew Cholmondeley, York Regional Police. All rights reserved.)

noisy. There are times when nothing can be done after the fact to remedy the background pattern. Could a subtraction image have been done at the point of capture? Perhaps, but that opportunity is lost in the past. Figure 7.5 is actually a fun one. Open in Photoshop, and scroll through the channels of RGB. Does anything stand out? Now convert to LAB color mode and scroll through the channels. How about now?

In RGB color mode, every channel contains noise, but in LAB color mode, there are some interesting things happening. The lightness channel contains all of the ridge information, as it has been dusted with black (colorless), and we know all neutral and contrast information lives in the L channel, but in the "a" channel, the colored flowers stand out brightly because they are magenta, and the greenery stands out darkly. The "b" channel is almost the inverse of the "a," and the magenta parts are dark, while the green parts are light. Meanwhile, the flowers and greenery are exhibiting as dark noise in the lightness channel.

The Strategy

Create a channel of noise, to subtract *from* the lightness channel. The goal is now to combine the lightest parts of the "a" channel (noise), with the lightest parts of the "b" channel (also noise) to create one channel containing all the noise. *That* channel of light noise will be *inverted* to make the noise a dark entity, and *then* it will be subtracted from the lightness channel. Here's how:

Create the noise channel:

- Open the image and convert to LAB color mode.
- Go to Calculations.
    - Source 1—"a" channel
    - Source 2—"b" channel
    - Blending mode—Lighten
    - Opacity—100%
    - Result—New channel (Alpha channel)

An alpha channel has been created that contains the noise from both the "a" and the "b." Why was the "lighten" blending mode used, and how does it work? Setting the blending mode to *Lighten* tells Calculations to compare every pixel address between the two channels and use the lightest one. The magenta flowers of the "a" channel were light, and the green foliage in the "b" channel was light. The background this pattern was printed on is a neutral white, so it is the same in both the "a" and the "b"—50% gray. The pattern of both the flowers and the foliage is now represented as light noise together in one channel. Give this channel a little contrast adjustment in levels before attempting the subtraction. Not too drastic, though a little trial and error here would be expected (my Levels settings: Black point—25, white point—237).

Perform the subtraction (Figure 7.6):

- Go to Calculations.
    - Source 1—Alpha channel (noise): Check the invert box to make the noise dark

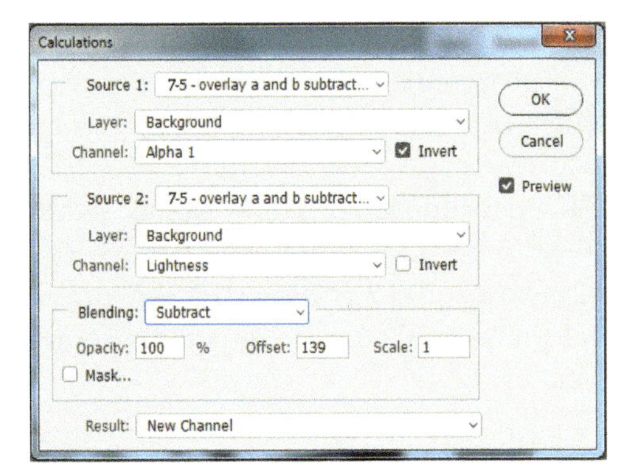

Figure 7.6 The Calculations dialogue box being used to subtract the noise channel just created, from the lightness channel.

Figure 7.7 (Right) The noise channel created by blending both the "a" channel and the "b" channel in the "lighten" blend mode. (left) The result of subtracting our noise channel from the lightness channel.

- Source 2—Lightness channel
- Blending mode—Subtraction
- Opacity—100% (compare between 50, 75, and 100%)
- Offset—Roughly between 100—140
- Result—New channel

The best values used for opacity and offset will be different for everyone, depending on the contrast adjustments made in the noise/alpha channel (Figure 7.7).

Convert the image to grayscale, and make some minor contrast adjustments to complete the processing of this file.

So, that is a bit of review. Feeling confident you can run with this information? Awesome!

Overwhelmed? Keep it simple. Search for signal. If there is noise obscuring the signal detail, search for the noise in a channel where it may (or may not) reside by itself. Try to blend channels to strengthen, and/or subtract noise channels from signal channels. Contrast adjustment comes last. Practice.

Review Questions

1. How will I know which tools to use for any given image requiring enhancement and analysis?
2. What are the four main general steps to image enhancement/processing?
3. Describe an instance where some forethought and proactive action at the photography stage may result in clearer detail capture.
4. When working with a very high contrast image of a fluorescing impression under laser light, what file format has the potential to glean the best detail in shadows and highlights?
5. Why is this so?

Digital Imaging in the Courts

8

Problem solving is hunting. It is savage pleasure, and we are born to it.

Thomas Harris

In this final chapter, the acceptance of testimony based on digital image processing is examined on a selected case basis. This is far from a complete list of cases featuring the successful introduction of digital evidence. It is based on the experience and the research of the authors and colleagues. As might be expected, some citations have more data associated with them than others.

Issues of tendering conventional fingerprint evidence lie outside the purview of this book. The focus here is on the process of creating an optimum digital image upon which a trained professional can execute the actions of analysis, comparison, evaluation, and verification (ACE-V), form an opinion, and defend that process in court. The forensic discipline is understandably conservative about adopting new technologies and procedures. The repercussions of accepting unproven technology with undue haste and unsupported conjecture are far greater than in other fields of endeavor.

Using new technologies to detect more evidence is one thing. Tendering that evidence in court is quite another. We have witnessed a stiffening of court standards for forensic evidence in recent years [1,2]. The phrase "junk science" has emerged to challenge our skill base and operations, and to refute acceptance and trust. Before we ask a court to accept our evidence, we must have a thorough understanding of what we have done, how we've done it, and why we've done it.

Best Evidence Rule

The Best Evidence Rule in Canadian law is a British common law rule of evidence, and it requires the proponent of evidence to produce the best evidence available to that party. It states, "*The law does not permit a man to give evidence which from its very nature shows that there is better evidence within his reach, which he does not produce*" [3]. The spirit of this dictum states that we have an *obligation* to provide the court with the clearest, truest, and most complete information and images that it is possible for us to prepare. Not only are we permitted to use new proven technology in pursuit of best evidence, we have an obligation to do so. It is not an imagination stretch in any jurisdiction that forensic professionals are expected to recover and present in court the clearest and most informative impression evidence possible.

This rule is frequently associated with the requirement to produce original documents rather than facsimiles, but it reaches back to the eighteenth century, long before photocopy and facsimile reproduction. It seems to be unmistakable and patently obvious in its intent to discourage and disallow the "low-hanging fruit" proclivity, which has negative consequences for any undertaking. Nowhere are the outcomes of this half-hearted approach more serious than in forensic science.

Diagnosis

Clearly, there is a diagnostic element inherent to evidence recording, which is an essential prerequisite of the ACE-V process: analysis, comparison, evaluation—verification. *We identify a visible entity as desirable forensic signal and everything that interferes with it as noise.*

Before the issue ever arises as to whose fingerprint it is, the determination must be made that it is, in fact, ridge detail, and that it was not somehow created in whole or in part by some unknown physical phenomenon. This is an essential part of the process each time a fingerprint technician begins examination of a fingerprint, regardless of whether he/she developed it or discovered it as visible.

Development of ridge detail by the dusting method relies on the ridge deposit being sticky, while the substrate is not. Fingerprint technicians are aware that other unknown oils and moisture not deposited by ridge detail can also retain powder, frequently abutting or overlaying areas of ridge detail.

It is common knowledge amongst practitioners that ninhydrin (for example) is a reagent may develop areas on an exhibit that were not introduced by finger contact, but are occasionally on or part of the same site as the fingerprint. Drops of perspiration containing the compounds targeted by ninhydrin or other chemistry can land and *have* landed, abutting or even on top of a fingerprint, obscuring that portion of it. Superimposed areas of ridge detail are routinely encountered on criminal exhibits.

Despite these occurrences, we can recognize areas of continuous, reliable ridge detail, upon which we can base an opinion. Identification professionals are trained to recognize an area of significant pattern evidence for what it is, be it deposited by hand or foot, separate from artifacts and entities of other sources. Having made the diagnosis that the subject is a viable evidence impression (signal), optimized and prepared it for analysis and comparison to a known standard, there are three questions that require answers.

What Have We Done?

A familiarity with the precise meaning of words that have been used in court proceedings is also important. The following definitions are excerpts from Webster's Dictionary (relevant portions) [4].

Alter

- *To make different without changing into something else*

Change

- *To make different in some particular*

Enhance

- *To increase or improve in value, quality, desirability, or attractiveness*

Manipulate

- *To treat or operate with the hands or by mechanical means especially in a skillful manner*
- *To manage or utilize skillfully b: to control or play upon by artful, unfair, or insidious means especially to one's own advantage*
- *To change by artful or unfair means so as to serve one's purpose*

Restore

- *To put or bring back into existence or use*
- *To bring back to or put back into a former or original state*

These terms have been used in court proceedings at various times to describe the processes of obtaining the best possible image of evidence upon which to conduct the ACE-V process. Some have been suggested in an adversarial context, as part of cross-examination (manipulation, for example). It is not in the best interest of accuracy to accept an inaccurate definition or an incomplete one.

Most accurately, we have identified through diagnosis, the forensic signal in an image and have, to the best of our ability, optimized the signal-to-noise ratio.

Why Have We Done It?

We have done it in compliance with the spirit of the Best Evidence Rule.

How Have We Done It?

We have employed the skills and the technologies used and accepted within the discipline of forensic identification. Digital techniques can often take up where analog techniques fall short in this optimization of forensic signal. The optimization process from diagnosis to final image may involve chemistry, special lighting, selective filtering, conventional photography, digital image processing, or a combination of these elements.

Analog Optimization of Images

Image optimization was not introduced with digital technology. Evidence photographers have been practicing analog image processing since the dawn of photography.

Some analog techniques employed for film image optimization include:

- Contrast adjustment (through choice of film, developer, developing time and printing paper grade)
- Contrast filters
- Painting with light
- Angle of light
- Polarizing filters
- Narrow band filters
- Selective excitation wavelength
- Selective focus
- Unsharp mask

Evidence photography is not simply or exclusively a process of recording what we can easily see. There is a necessary process involving not only technical ability but also reason, judgment, and expert discernment:

- If a night scene must be documented, painting with light effectively reveals previously unseen detail, but it is not a normal recording of the scene as anyone remembered it.
- Ninhydrin fingerprints photographed with a green filter often reveal far more ridge detail than was apparent without it.
- Infrared luminescence can be used to record evidence that cannot be seen at all by the human eye.
- Polarized light examination reveals striations in plastic bags (unseen in conventional lighting) that may be used to connect items of common manufacture or to link heat-seal signatures to a specific machine.

- Unsharp mask is a technique used historically in the analog domain by astronomers to record edge detail of bright objects in space. It has been adapted for use in the digital domain.

Digital Optimization

Simply put, we use one part of our training to recognize forensic signal (fingerprint, for example) and another part to make a recording of it as distinct and separate from interfering background as possible.

Digital enhancement may include, but is not limited to:

- Spatial filters
- Channel blending
- Dynamic range adjustment (contrast)
- Image subtraction
- Fast Fourier transform
- Multiframe averaging (video)

Testability

It is patently obvious that if we ask a court to believe, and believe in the technology we use to detect evidence, we had better believe in it ourselves.

The question has been asked, *"Could the application of computer techniques to a digital image result in artifacts that could mislead or possibly even miscarry justice to the disadvantage of the accused?"* This question addresses only random occurrence or inadvertence, not malfeasance. An individual does not require the use of a computer to be dishonest.

First, the level two details upon which we largely base our conclusions are not binary point spots but terminations of whole ridges or creations of new ones where they bifurcate. They constitute an unbroken chain of agreement in sequence. Adding artifacts to blend with an existing ridge detail impression is not comparable to adding "1" to an existing "4" to make an indistinguishable "5". Such artifacts would have to be the whole ridge or at least the portion of it present in the fingerprint. They would have to closely resemble true ridge detail in appearance. They would also have to occur exactly at positions of false agreement with the known print and *not* anywhere else, blending exactly with the bona fide ridge detail in the fingerprint to falsely match the fingerprint of a suspect under scrutiny.

This scenario is reminiscent of the analogy in which a hurricane blows through a junkyard and builds a jet aircraft, when the reality is the reverse. Even if such minutiae-emulating artifacts *were* inherent to the digital enhancement routines, their randomness could only work to the advantage of an accused person.

Second, we can establish the degree to which our technology produces artifacts or "false positives" by repeatedly testing the process on known fingerprints we have intentionally obscured. We can then compare the final processed results to the original impressions. Conducting many such comparisons under controlled conditions gives one a high level of confidence in assessing the case applications. Only uncompressed file formats should be used to record impression evidence.

It has been the writer's experience in assessing a large volume of such tests that no such artifacts were noted that were attributable to the digital processes applied, nor was there any added presence that could in any way be mistaken for ridge detail. In short, the differences were in appearance only, comparable to the differences observed between two fingerprints made by the same digit but recorded in different media or on different surfaces.

At the 1997 International Association for Identification's (IAI) 82nd Annual Training Conference in Boston, Massachusetts, the IAI passed Resolution 97-9, which recognized digital imaging as a "scientifically valid and proven technology" for the first time [5].

Research

The Internet is the most powerful tool we possess for compiling information. In this information age, unlike any other time in history, advances in science and technology may be shared instantaneously across the world, and forensic digital processing technology is no exception; image processing techniques, new software and image applications, court citations, research papers, legal precedents, and expert correspondence are all readily available to the forensic investigator today.

Digital Image Processing—Introduction into Court

In each of these examples, science and the law have relied heavily on the information recorded in digital images.

1991—*U.S. v. Knight*

- Homicide
- Fingerprint enhanced by FFT
- Challenged by defense
- Subject of Frye hearing
- Evidence upheld
- Four life sentences [6] (Figure 8.1)

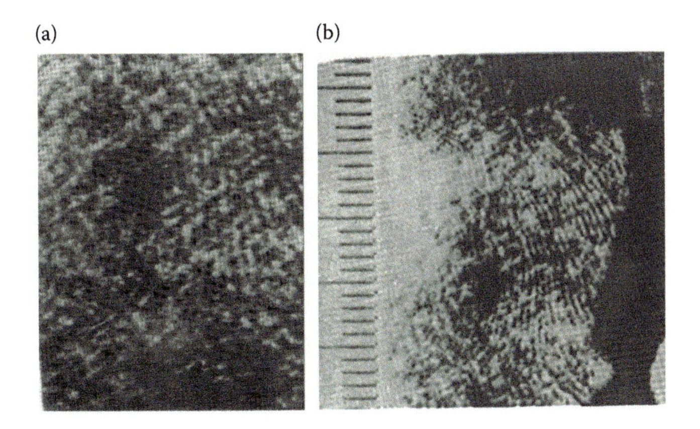

Figure 8.1 (a) Fingerprint on a pillowcase after DFO processing. (b) Fingerprint after FFT editing to remove weave pattern. (Courtesy of Henrico County Police Department.)

1991—Canada

- Perjury, obstruct justice, police officer charged
- Techniques used
    - Line histogram
    - Edge and transition adjustment
    - Contrast adjustment
- Evidence tendered and unchallenged

1993—*R. v. Gough/Alexander* (Canada)

- Homicide, three defendants
- Store clerk killed with shotgun
- Videotape multiframe averaging
- Shoeprint impression developed with powder on countertop
- Lift attempt unsuccessful
- Re-treated with different fingerprint powder
- Second lift attempt unsuccessful
- Erasure subtraction conducted on impression
- Size, make, and model of shoe determined from result
- Shoeprint was key evidence in determining shooter
- Evidence tendered, unchallenged [7,8] (Figure 8.2)

1993—Reginald Denny Beating

- Enhanced photos and videotape
- Instrumental in suspect identification
- Ruled admissible by Superior Court [9–11]

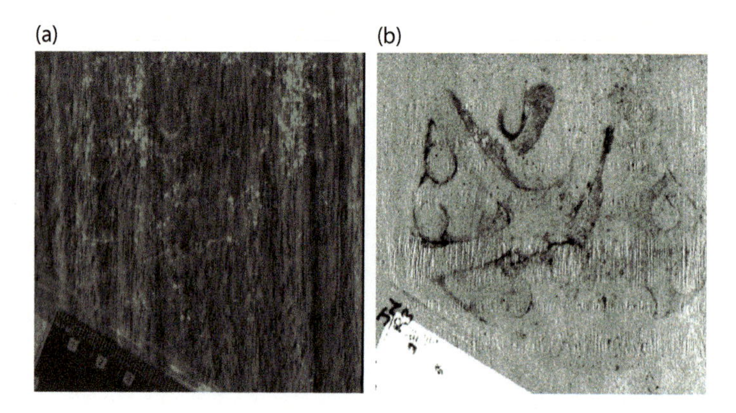

Figure 8.2 Gough/Alexander. (a) Woodgrain countertop after two unsuccessful powder and lift attempts. (b) Shoeprint after erasure subtraction conducted. (Courtesy of Queen's Printer for Ontario 2017. Reproduced with permission. All rights reserved.)

1995—*U.S. v. Hayden*

- Homicide
- Bloody finger and palm prints on bedsheet
- Suspect identified after FFT pattern removal
- Challenged by defense
- Subject of Frye hearing
- Conviction affirmed [12,13] (Figure 8.3)

1995—*California v. Phillip Lee Jackson*

- Double murder
- Fingerprint digitally enhanced and identified
- Frye hearing requested by defense
- Ruled unnecessary by court, as digital image processing is a readily accepted procedure [13]

1996—*R. v. Cooper*

- Homicide
- Faint blue ballpoint writing on badly creased cardboard key tag
- Second image recorded with blue filter (Kodak Wratten 47)
- Enhanced through filtration subtraction
- Revealed previously indecipherable writing
- Led to finding of new evidence crucial to investigation
- Evidence tendered unchallenged (Figure 8.4)

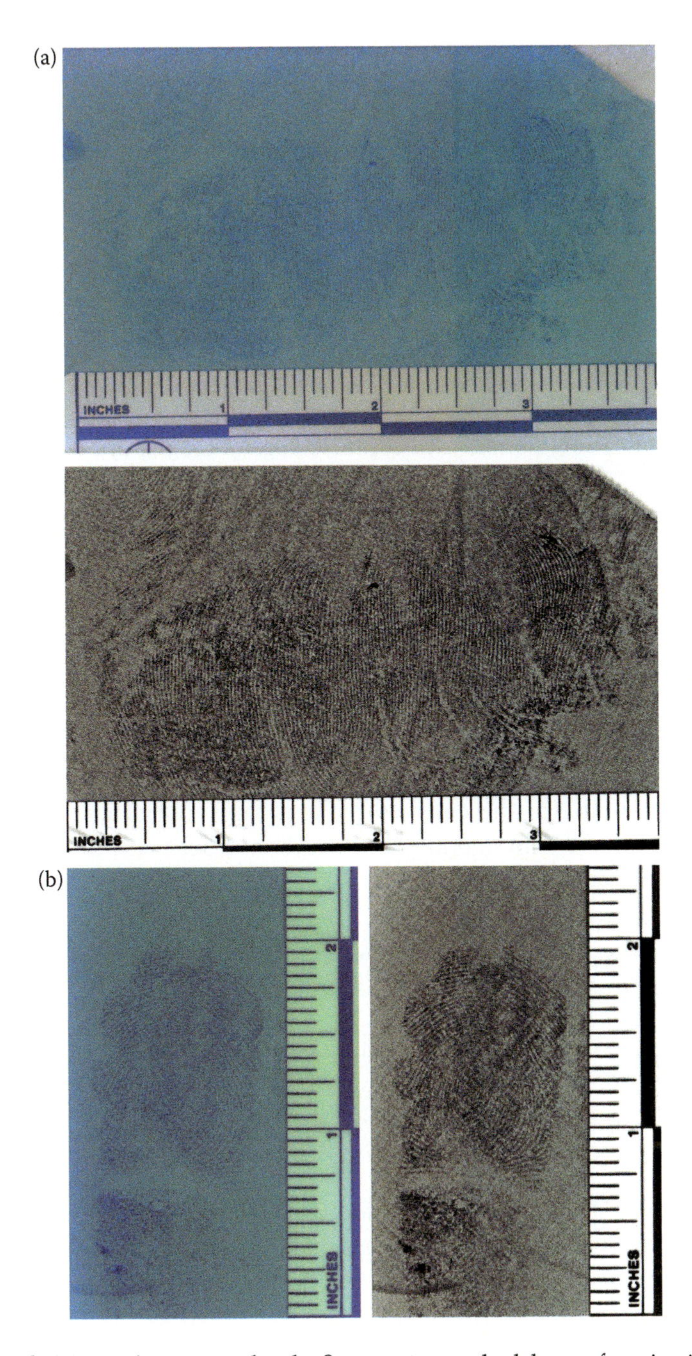

Figure 8.3 (a) Hayden. Top: Bloody fingerprint on bedsheet after Amido black processing. Bottom: Fingerprint after FFT editing to remove fabric weave pattern. (Courtesy of Erik Berg.) (b) Hayden. Left: Bloody fingerprint on bedsheet after Amido black processing. Right: Fingerprint after FFT editing to remove fabric weave pattern. (Courtesy of Erik Berg. All rights reserved.)

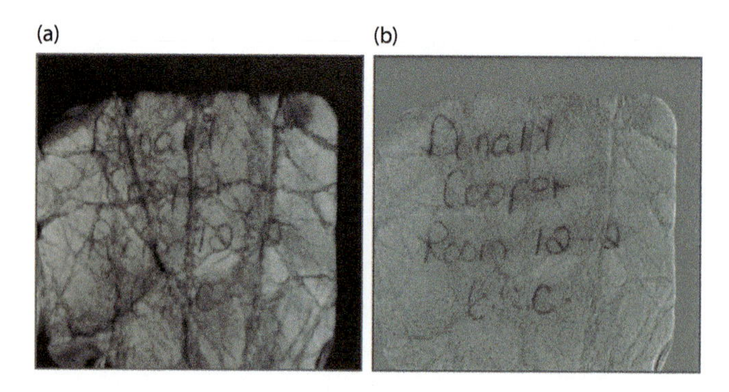

Figure 8.4 Cooper. (a) Blue ballpoint writing on badly creased cardboard key tag. (b) Result of subtraction with second image photographed with blue filter (Kodak Wratten 47). (Courtesy of Queen's Printer for Ontario 2017. Reproduced with permission. All rights reserved.)

1998—*The State of Ohio v. Brett X. Hartman*

- Homicide
- Digitally enhanced fingerprints
- Hartman sentenced to death
- Conviction appealed in 2001, challenging
  - The admission of digitally enhanced fingerprints
  - The failure of the court to make a threshold determination of the expert who conducted the enhancement and identification processes
- Court of Appeals ruled that
  - Computer enhancement of fingerprints was appropriate
  - The fingerprint witness was qualified to identify the defendant's fingerprint
- Defendant put to death by lethal injection in 2012 [5]

1999—*R. v. Gill* (British Columbia, Canada)

- Robbery
- Videotape digitized and enhanced
- Challenged by defense as inadmissible
- Ruled by court as admissible and relevant
- Guilty as charged [14]

2001—*Florida v. Victor Reyes*

- Duct tape located on wrappings of victim's body
- Fingerprint developed on duct tape were digitally enhanced and subsequently identified as made by the accused

- Image processing evidence tendered in court
- Challenged by defense in Frye hearing
- Court ruled that digital enhancement of images is an accepted process throughout the forensic community
- Victor Reyes acquitted of first-degree murder by jury
- Defense referred to digital technology as "junk science" and asserted that jury had rejected it as such
- Post-trial, members of jury stated their acceptance
  - Of the digital technology introduced in the trial
  - Of the identification of the fingerprint as Victor Reyes
  - They had based their verdict on the following:
    - Last-minute no-show of key witness
    - Location of fingerprint on duct tape did not necessarily establish Reyes as the murderer [5,14–16]

2001—*Mastro Almond v. State of Georgia*

- Georgia Supreme Court (GSC)
  - Defense argued that
    - Trial court erred in admitting photographs captured with a digital camera
  - GSC ruled that
    - Pictures had been properly authenticated by the prosecution as fair and truthful representations of what they purported to depict
    - No reason to differentiate in admitting images from a digital camera and a conventional film camera [5]

2003—*People v. Perez*

- Computer-enhanced footwear impressions admitted into evidence
- Challenged by defense on the grounds that admission of said evidence was in error because:
  - New scientific procedures were employed
  - No Frye-Kelly hearing was held
- Court ruled that
  - Adobe Photoshop was not a new or novel technique
  - Digital enhancement is generally accepted by the scientific community as reliable [5]

References

1. Daubert Standard, Cornell Law School, available at https://www.law.cornell.edu/wex/daubert_standard

2. Mohan Criteria, Irwin Law, available at https://www.irwinlaw.com/cold/mohan_criteria
3. Canada Evidence Act, available at http://laws-lois.justice.gc.ca/eng/acts/C-5/
4. Merriam-Webster, available at https://www.merriam-webster.com
5. D. Witzke, personal communication, 2017.
6. The Forensics Library, available at http://aboutforensics.co.uk/robert-knight/
7. *Regina v. Gough/Alexander*, 1993, 23-Feb-93, 2872 Ellesmere Road, 27,Horacio, Diogo, Male, shooting, Sherwin Clinton Alexander and Hugh Silvinas Gough convicted in 1995, available at http://maps.library.utoronto.ca/datapub/toronto/homicide/toronto-homicides_1990-2013.csv.bak
8. J. Norman, personal communication, 2016.
9. L. Rudin and S. Bramble, *Investigative Image Processing*, 2942, National Institute of Standards and Technology (U.S.), National Institute of Justice (U.S.) SPIE, 1997—Law enforcement.
10. M. King, 2017, *When Riot Cops Are Not Enough*, p. 219, available at https://books.google.ca/books?isbn=0813583764
11. K. Devlin and G. Lorden, *The Numbers behind NUMB3RS: Solving Crime with Mathematics*, pp. 66–68, available at https://books.google.ca/books?isbn=0452288576
12. FindLaw, *State v. Hayden*, available at http://caselaw.findlaw.com/wa-court-of-appeals/1189825.html
13. S. Staggs, The admissibility of digital photographs in court, available at http://www.crime-scene-investigator.net/admissibilityofdigital.html
14. E. Neate, Digital images as evidence, available at http://www.neateimaging.com/evidence.PDF
15. Circuit Court, 17th Judicial Circuit, Broward County, Florida, available at http://www.forensictv.net/Downloads/legal/florida_v._reyes_digital_enhancement.pdf
16. E. Berg, personal communication, 2016–2017.
17. D. Witzke, It's a photo finish, not junk science, available at https://www.foray.com/images/pdfs/JUNKSCIENCEREBUTTAL-1.pdf

Further Reading

Adobe Resource Library: Color Management: The color conundrum Adobe Help files, available at https://helpx.adobe.com/lightroom/help/color-management.html

M. J. Langford, *Basic Photography*, Third Edition, Focal Press, London and New York, 1973.

B. Long, *Complete Digital Photography*, Seventh Edition, Course Technology Cengage Learning, USA, 2013.

R. D. Olsen, Sr., *Scott's Fingerprint Mechanics*, Charles C Thomas, Illinois, 1978.

G. Reis, *Photoshop CS3 for Forensic Professionals*, Symbex—Wiley, Indianapolis, 2007.

J. C. Russ, *Forensic Uses of Digital Imaging*, CRC Press, Boca Raton, FL, 2001.

E. M. Robinson, *Crime Scene Photography*, Third Edition, Academic Press, London, 2016.

J. Smith, Image enhancement and Adobe Photoshop: Using calculations to extract image detail. *Journal of Forensic Identification* 57(4), July/August 2007.

J. Smith, Computer fingerprint enhancement: The joy of LAB color. *Journal of Forensic Identification* 62(5), September/October 2012.

Index

D

E

G

H

I

J

K

L

M

R

S

T

U

V

W

X

Z